A COMPLETE GUIDE TO
MONKEYS, APES AND OTHER PRIMATES

A COMPLETE GUIDE TO
MONKEYS, APES AND OTHER PRIMATES

Michael Kavanagh · Introduced by Desmond Morris

JONATHAN CAPE
THIRTY BEDFORD SQUARE LONDON

To my father,
to Aren't
and
to the memory of my mother

This book was designed and produced by
The Oregon Press Limited,
Faraday House, 8–10 Charing Cross Road,
London WC2H OHG

First published in 1983 by
Jonathan Cape Limited, 30 Bedford
Square, London WC1

British Library Cataloguing in Publication Data

Kavanagh, Michael
A complete guide to monkeys, apes and other primates.
1. Primates
I. Title
599.8 QL737.P9

ISBN 0-224-02168-0

Design: Gail Engert
Reader: Raymond Kaye
Maps: The Kirkham Studioes, Midhurst, West Sussex

Typeset by SX Composing Ltd, Rayleigh, Essex
Printed and bound by New Interlitho SpA,
Milan, Italy

FRONTISPIECE **The female vervet monkey's nipples are so close together that her infant can suck on both at once**

CONTENTS

ACKNOWLEDGMENTS

In telling the primates' story, I have tried to avoid too much mention of detailed complications or exceptions to rules and to concentrate on the main points of interest about the different species. Primates are wild animals so it is hardly surprising that facts known about one species are often not known about another. Nevertheless, all the information that is presented has been carefully checked against the scientific literature. I have, of course, drawn upon my own research in zoos, in laboratories and in the wild parts of South America, Africa and Asia, but most of what follows has been gleaned from the talks, books and papers of my primatological colleagues.

In the world of science, we draw freely upon each other's published work and it is a rule that we always state our sources. We do this partly so that others may check that we have not made mistakes, and partly as a common courtesy. Writing a book such as this has presented me with something of a dilemma. On the one hand, I wish to acknowledge those whose efforts form the basis of the text, but on the other hand I have not wanted to disrupt the flow of the words with hundreds of references. I hope that my colleagues will accept as adequate a list of those who, generally unwittingly, have contributed to this guide:

M. L. Allen, J. & S. A. Altmann, J. M. Ayres, J. D. & J. I. Baldwin, S. K. Bearder, E. L. Bennett, C. Bonvinvino, D. B. Buchanan, G. Buchli, N. Budnitz, J. Buettner-Janusch, J. O. Caldecott, C. R. Carpenter, M. Cartmill, R. Castro, N. R. Chalmers, P. Charles-Dominique, D. J. Chivers, T. H. J. Clutton-Brock, A. F. Coimbra-Filho, R. W. Cooper, J. H. Crook, R. Dare, K. Dainis, I. Davidson, A. G. Davies. G. W. H. Davison, G. A. Dawson, J. M. Deag, I. DeVore, W. P. J. Dittus, G. A. Doyle, F. V. DuMond, R. I. M. Dunbar, J. F. Eisenberg, J. Fa, J. G. Fleagle, M. P. L. Fogden, R. Fontaine, J. Fooden, D. Fossey, R. S. Fouts, C. H. Freese, G. Galat, A. Galat-Luong, B. M. F. Galdikas, B. T. & R. A. Gardner, J. S. Gartlan, A. Gautier-Hion, S. P. Gittins, A. G. Goodall, J. Goodall, C. P. Groves, M. Guerreiro de Lima, K. R. L. Hall, D. A. Hamburg, A. H. Harcourt, C. S. Harcourt, R. S. O. Harding, J. E. Harrington, P. G. Heltne, J. Hernandez-Camacho, P. Hershkovitz, C. A. Hill, W. C. O. Hill, A. & C. M. Hladik, K. M. Homewood, S. B. Hrdy, J. Ingram, T. Inskipp, K. Izawa, C. H. Janson, A. Jolly, P. Jouventin, M. Kawai, S. Kawamura, J. Kingdon, W. G. Kinzey, D. J. & L. L. Klein, H. Kummer, L. Leland, W. B. Lemmon, L. K. Lippold, J. MacKinnon, A. Magnanini, G. H. Manley, P. R. Marler, E. R. & J. T. Marshall, R. D. Martin, E. R. McCowan, W. C. McGrew, D. McKey, F. G. Merfield, K. Milton, R. A. Mittermeier, S. M. Mohnot, D. & R. Morris, M. Moynihan, R. D. Nadler, J. R. & P. H. Napier, P. F. Neyman, C. Niemitz, N. Nightingale, T. Nishida, J. F. Oates, J. M. S. Oliveira, J. R. Oppenheimer, E. Pages, J. J. Petter, A. Petter-Rousseaux, A. Peyrieras, J. I. Pollock, A. G. Pook, D. Premack, A. Pusey, J. J. Raemaekers, M. F. Ramirez, N. L. Rettig, J. Revilla, A. F. Richard, H. D. Rijksen, D. Roberts, W. A. Rodgers, M. L. Roonwal, K. D. Rose, H. Rothe, T. E. Rowell, R. Rudran, D. M. Rumbaugh, A. Schilling, C. Schürmann, C. Smith, P. Soini, E. D. Starin, T. T. Struhsaker, S. Strum, Y. Sugiyuma, R. W. Sussman, A. Suzuki, I. Tattersall, R. L. Tilson, C. Torres de Assumpção, C. E. G. Tutin, M. G. M. van Roosmalen, S. Vellayan, S. J. Wallis, M. S. & P. Waser, P. G. Waterman, S. Wells, A. J. Whitten, J. F. Wojcik, K. Wolf, R. W. Wrangham, P. C. Wright, Y. Z. Zhang, and various biologists connected with the Gombe Stream Research Centre in Tanzania, Karisoke Research Station in Rwanda, Kibale Forest Project in Uganda and the Malaysian Primates Research Programme.

In addition, advice, criticism and other kinds of support have been generously given by E. B. M. Barrett, S. K. Bearder, E. L. Bennett, J. O. Caldecott, P. Chai, A. G. Davies, G. W. H. Davidson, A. G. Dixson, R. I. M. Dunbar, J. Fa, P. H. Harvey, A. D. Johns, N. M. Kavanagh, S. Kavanagh, L. Leland, J. J. Petter, L. Stewart, P. M. Stewart, T. T. Struhsaker, C. Torres de Assumpção and especially José Marcio Ayres. The typing was done quickly and efficiently by Susan Smith, to whom I am also very grateful.

Above all, I must thank Michael Rainbird and Mary Anne Sanders of The Oregon Press for asking me to write the book in the first place, for working so hard on it with me and for continuing to believe in it throughout all the stages of its production, and Joëlle Patient for, among many other things, keeping me reasonably sane while it obsessed me.

MICHAEL KAVANAGH
Kuching, Sarawak
March 1983

This is the first book on primates to include a complete set of photographs covering every living genus of the monkeys, apes and their relatives. This unique feature alone makes it indispensable to anyone interested in these fascinating animals.

The book also contains up-to-date details on each genus, providing much new information since the last general survey – *A Handbook of the Living Primates*, by J. R. and P. R. Napier – was published, as long ago as 1967. A great deal of research and fieldwork has been carried out on primates since the 1960s and it was high time that someone incorporated the results of these recent investigations in a new volume.

It is especially important that we should examine where the various kinds of primates live today, and this the book facilitates with a series of useful distribution maps. Since most members of the order are found in forests, the future of the group is seriously threatened by the ever-increasing destruction of their habitat. No matter how well we protect the monkeys and apes themselves from hunting and exploitation, they will soon be homseless unless the worldwide chainsaw massacre can be halted. A few open-country and rock-dwelling species may survive, but the rest are destined to become hothouse zoo curiosities. If urgent action is not taken to protect the remaining forests, a new edition of this book published in, say, fifty years time, will be a very slim volume indeed.

The problem, of course, stems from the unparalleled success of one species of primate at the expense of all the others – the species I once called the naked ape. Other scientists referred to the species as 'man', but I wanted to remind my readers as vividly as possible of our membership of the primate order. In my opinion, it is the only way in which we can truly understand ourselves. If we concentrate instead on the uniquely human qualities of technology, language, philosophy and cultural variability, we soon forget our humble origins and start to weave fantasies about our immunity from biological laws. We stress how clever and adaptable we are – how we can do anything, adapt to any environment, survive anywhere. There is a terrible danger in this and we are already in a perilous condition. Every day we add another 150,000 to our world population of teeming humanity – a daily figure which exceeds the total world population of most of the other primate species discussed in this book. We are breeding so fast that we are swamping out everything else on the planet. We are becoming an infestation, encouraged by power-hungry politicians and purblind pontiffs.

If we were really as adaptable as some social scientists would have us believe, then it would not matter too much. We could live in a synthetic machine-world on land-masses totally covered by 1000-storey buildings, like some giant form of social insect. But that is simply not the sort of animal we are. We are primates, first and foremost, and if we are over-crowded in unnatural conditions we will destroy one another, just like any other member of the order under similar circumstances. We have a set of primate needs and we cannot shed them in a mere few thousand years.

Before we reached the stage of destroying one another, however, we would first eliminate all our non-human competitors for global standing-room. We would develop synthetic foods – not too daunting a task – and then set about systematically exterminating all natural forms of non-human life, including all animals and plants. All diseases would go too, with medical progress. Then we would build and build and disappear inside seething continent-sized edifices – the most successful animal in the world – and the winner of the greatest Pyrrhic victory in the story of our planet.

Before long the whole complex system would start to break down. Our biological, primate qualities would assert themselves and, as we became ever more cramped and crowded, we would suffer from intolerable stresses leading to acts of violence and disruption on an unimaginable scale. Our behaviour patterns would fragment. Our whole species would begin to behave as if mad.

If this sounds like pure science fiction, I can only assume that you have been fortunate enough never to have visited present-day humanity in its worst and most degradingly crowded city slums. There, the process I have described can already be observed in embryo. It is not something of the far-distant future, it is already happening today.

The alternative is to admit now, without delay, that we are not fallen angels, somehow protected by a divine force, but risen apes, susceptible, like any other animal, to environmental deprivation. Like any monkey or ape, we have certain fundamental requirements, with respect to living space, social relationships, family stability, group membership, face-to-face communication and sexual interaction. Like them, we have a complicated body language, a set of subtle facial expressions, an intricate physiology demanding a balanced, omnivorous diet, a system of sex hormones, a pattern of intense parental responses and a high level of curiosity.

Monkeys and apes have fascinated humans and been among their objects of religious veneration for thousands of years. This holy man of India shows great respect for the sacred langur.

These are all biological features which we share with every other species of higher primate.

We *are* flexible, it is true, but our flexibility is much more superficial than we like to think. We may wear funny hats that make us look different at a distance, but beneath the wild variety of headgear the 4600 million human heads (and there will be an estimated 6000 million by the year 2000) are all buzzing with much the same emotions and desires and frustrations.

This fact makes it particularly important that we should study closely the other primates. Their similarities to us and their differences from us can teach us many salutary lessons. The more we can come to accept ourselves as members of their family, the safer we will be. We might even be able to face the plain truth that there are far too many of us for a primate species and do something to ease off the human breeding rate. If we respond intelligently to the present dilemma, we would not only be able to enjoy one another's company more, but we could be more generous in allowing other species to share the limited land mass with us. We could even attempt to reverse the process of destruction of the wild habitats, and work towards their positive increase once more. We would then be able to savour the richly varied environment that is part of our animal heritage.

In particular, we could take delight in the spectacle of our closest living relatives in their natural world. There was a time, of course, when monkeys were looked upon as dirty, filthy and obscene, and apes as either funny little clowns or great, hairy, raping monsters. As an early English author put it, in the twelfth century, 'Admitting that the whole of the monkey is disgraceful, yet their bottoms really are exceptionally disgraceful and horrible'. As late as the Reformation, Martin Luther used the terms 'ape' and 'devil' interchangeably. The great Thomas Jefferson believed that apes made a habit of raping negresses. Little wonder that when Darwin suggested a family relationship between man and monkey, he was abused and ridiculed. The public of his day was simply not aware of the beauty, complexity and variety of primates in their natural setting. They had not listened to a gibbon chorus at dawn, watched gorillas gently grooming one another's fur, marvelled at the acrobatic grace of a forest-canopy monkey as it sailed through the air, or the parental devotion of a tiny, heavily-burdened marmoset carrying its young; nor had they witnessed a gang of male baboons bravely challenging a marauding leopard, or a gathering of chimpanzees performing their spectacular rain-dance. The real world of Primates was virtually unknown to them. But there can be no such excuses now. Travel, films, television programmes – and books like the present one – make it impossible for all but the most obtuse of us to see monkeys and apes in the old light. With our new knowledge, we should be proud to be primates, rather than ashamed of the fact. Study the pages of this book and you will see what I mean.

DESMOND MORRIS
Oxford, 1983

The classification of living things

This is a book about our living relatives, the primates. With many of them, the kinship is obvious, but some seem at first to show few of the family characteristics. To understand why we must accept them in our order, we must first understand the rules by which these things are decided.

There may be as many as ten million living species of plants and animals in the world, most of which have yet to be described. Without some sort of organization, biologists would have a hopeless task in trying to remember them at all, let alone in trying to comprehend the relationships and interactions between them. The popular names of animals and plants often give some sort of a guide to relationships, but unfortunately this is inadequate. For example, yellow-tailed woolly monkeys and Humboldt's woolly monkeys might seem to be related from their names; but are they really related in the sense of being kin, or is it just that they both have a rather sheeplike look? And how do they relate to spider monkeys or even to woolly spider monkeys? And all this is in English; where does it get us to use Latin American names like *barrigudo*, *mono* and *muriqui*, as happens where these monkeys are found?

To add to the confusion, since there are no rules about the use of popular names, the same ones may be applied to different animals. There are, for example, three different 'golden monkeys' in the world, two in Asia and one in Africa. They are all roughly the same colour, but apart from that, they are neither particularly alike nor particularly closely related. I have tried, in this book, to avoid using the more confusing of popular names; but more importantly, 'taxonomy', or the science of classifying living things, is available to bring order out of chaos.

Modern taxonomy was started in the mid-eighteenth century by a brilliant and indefatigable Swede called Carolus Linnaeus (1707–78). He decided that all organisms should be classified according to a descending series of categories, the most important of which were to be expressed by two names. The first of these gives the 'genus' (plural 'genera') and the second the 'species' (plural also 'species'). The system is now universally used by biologists.

Neither species nor genus is very easy to define with precision, but the task was made easier for Linnaeus by his belief that each type of organism had arisen as an act of God, that it did not change over time and that species could be defined in terms of physical similarities. Modern science ignores the possible role of any deity and takes the view that species arise slowly in time as a result of evolution, that they do therefore change over time and that there may be considerable physical dissimilarities between members of the same species. The crucial factor in determining whether or not any two animals belong to the same species is that there is no barrier to inter-breeding links between them unless different populations are geographically separated from each other.

Nature does not, of course, always follow such discrete categories, or at least not in an obvious way, so there is often argument about what is and what is not a recognizable division. Put simply, under natural circumstances, different species either cannot or do not normally interbreed, but in cases where they do so, no line of descendants is left. Horses and donkeys are separate species and they can mate only to produce sterile mules. On the other hand, different breeds of domestic dogs often bear little resemblance to each other, but even if a union between a chihuahua and an Irish wolfhound would present difficulties, both could mate with intermediate-sized dogs; and in the long run, they could have a common descendant. Thus all domestic dogs belong to a single species.

Within individual species, animals from different places may differ from each other in some consistent way. These are described as being of separate 'subspecies' or 'races', terms that should only be used for different populations, not for individuals that happen to be unusual. Unfortunately, taxonomists sometimes get rather carried away in their work and name new races on the basis of absolutely trivial individual differences.

It is even more difficult to define a genus because the term fits no natural category. It is perhaps easiest to think of a genus as a grouping of very closely related species, none of which has as close a relative outside the genus. Some species have no such close relatives, but they are still given a generic name. Humans, for example, belong to the genus *Homo*, of which ours is the only living species.

In order to keep the classification system clear and mutually comprehensible, each species may have only one recognized two-part name or 'binomial'. With few exceptions, this is the name that was first published in a description of an example of the organism concerned, an example that must be deposited in a public museum or herbarium, so that other scientists can go along and check the actual specimen. Thus arguments about conflicting scientific names are decided by priority; and there is even an International Commission of Nomenclature to adjudicate disputes.

All scientific names are expressed in Latin, or a Latinized form: many have been invented in modern times. *Homo* is the Latin word for a person, but not surprisingly, the ancient Romans did not have words for species that have been discovered since the fall of their empire some 1600 years ago. By convention, animals' generic and specific names are always either underlined or printed in *italics*, and the genus always begins with a capital letter whereas the species never does.

Scientists have a fairly free hand in inventing Latin names for newly discovered species. An animal or plant may be named after the person who discovered it, or perhaps in honour of a professor or other person eminent in his field of study. Alternatively, the local, popular name may be Latinized. Rather more usefully, the name may be descriptive as, for example, in *Saguinus bicolor* for a two-tone tamarin that will be described later on in this book. A certain amount of humour may be allowed, as in the case of an upright, cylindrical toadstool that is known to science as *Phallus impudicus* for unmistakable reasons. Offensive names are generally banned.

The names that express relationships above the generic level are also formally expressed in Latin, although they are not italicized. They give rise to even more controversy than the definitions of genera, but are nevertheless helpful in providing a framework within which to fit the species; and there tends to be more agreement than disagreement about the categories. The application of this system to the primates is shown in the taxonomic listing in the Appendix at the end of this book, p. 215, but it might help to see the categories into which four well-known primates fall.

	Human	**Common chimpanzee**	**Sacred langur**	**Lesser bushbaby**
Kingdom	Animalia	Animalia	Animalia	Animalia
Phylum	Chordata	Chordata	Chordata	Chordata
Class	Mammalia	Mammalia	Mammalia	Mammalia
Order	Primates	Primates	Primates	Primates
Superfamily	Hominoidea	Hominoidea	Cercopithecoidea	Lorisoidea
Family	Hominidae	Pongidae	Cercopithecidae	Lorisidae
Subfamily	Homininae	Ponginae	Colobinae	Galaginae
Genus	*Homo*	*Pan*	*Presbytis*	*Galago*
Species	*sapiens*	*troglodytes*	*entellus*	*senegalensis*

If these categories are not enough, others such as suborders, infra-orders, legions, tribes and cohorts are available.

The evolutionary process

Contrary to pre-Darwinian belief, the primates did not arise on this planet by some supernatural, instant process. They arose gradually, over millions of years, by means of the natural sequence that we know as evolution.

The most important mechanism by which evolution takes place was first described in 1858 in papers by Charles Darwin and Alfred Russel Wallace that were read into the records of a meeting of the Linnean Society in London. In a nutshell, their theory states the following:

1. there is a universal over-production of offspring by living organisms;
2. the environment cannot support all of these offspring, so there is a struggle for existence which will result in the demise of many; and
3. within a single species, even within the offspring of one set of parents, there is some variation between individuals and some of the characteristics that vary are inherited; therefore
4. individuals whose characteristics give them a competitive advantage over their fellows will be the ones to contribute more offspring to the next generation; and
5. each succeeding generation will contain a greater proportion of individuals that have the advantageous characteristics.

The process is summed up in the phrase 'the survival of the fittest'. Over a very long time, small changes between generations may give rise to big differences between distant ancestors and descendants. The mechanism of change is known as 'natural selection' and the characteristics that are selected are described as being 'adaptive'.

It is, however, important to realize that under stable environmental conditions, the evolutionary process will reach a point at which it favours stability in the characteristics of the species because each one will have become adapted to its way of life and most of the variations that arise will be disadvantageous. If a change in the environment then occurs, the species may no longer be so well adapted, giving a spur to the natural selection of new characteristics. This is why evolution appears to have proceeded faster during periods of climatic or other upheaval in the history of the world.

In reading about the many fascinating types of primates that are alive today, it is worth bearing in mind that behavioural as well as bodily characteristics can be inherited and that natural selection therefore works on behaviour as well as on physical form. Furthermore, it works on individuals rather than on groups or species, so forms of behaviour may well evolve that are not only selfish but positively disadvantageous to others. A simple example of this would be any case of selection for big, strong males that would compete fiercely to mate with as many females as possible. Clearly, under these conditions, one male's success is another's failure; and it is not the case, that individuals will 'go easy' on their rivals 'for the good of the species'.

The origin of the primates

Rather more than 250 million years ago, the first mammal-like creatures evolved from their reptilian forebears. Most of them were exceedingly small and for a long time they failed to make much of an impact. It is probably reasonable to assume that they remained rather shy and retiring and, after a mere 25 million years or so, they became dramatically overshadowed by a diverse array of gigantic reptiles. This was the Age of Dinosaurs.

Dinosaurs are aptly known as ruling reptiles, and rule they did for about 150 million years. Then, quite suddenly, they were gone. Of course, 'quite suddenly' is a relative expression and no doubt their demise was played out over a few million years; but it can be said that about 75 million years before the present, the dinosaurs were a spent force and the Age of Mammals was beginning.

It was not only the fauna of the Earth that was changing during this period. Massive geological and climatic upheavals were also taking place. Indeed, it must have been these physical influences that triggered off the numerous evolutionary chain reactions that began at about that time. This was when the flowering plants began their replacement of the more primitive non-flowering vegetation.

Amid all this change, the twenty modern orders of mammals began to appear, including four that probably had a common origin about 65 million years ago. Between them, these four orders gave rise to two huge success stories and two relative failures.

The first of the success stories is that of the Chiroptera, the bats. These animals are now virtually worldwide in distribution, they are the only mammals (other than humans) to have achieved powered flight, and they comprise the second largest mammalian order (after the rodents) in terms of having 18 families and more than 800 species. It is their ability to fly that lies behind their great success, since it has enabled them to exploit ways of life, or 'niches', that are closed to other mammals.

In contrast, the Dermoptera, or colugos, have to make do with gliding; and although they are very good at this, they have failed to proliferate on any great scale and the order is restricted to two species that live in the forests of South East Asia. Colugos are also known as flying lemurs but they are not lemurs and cannot fly.

The third of the four orders is the Scandentia, the tree shrews. This is the second of the relative failures, being restricted to a single Asian family of about 17 species. Tree shrews are rather squirrel-like creatures, although they may be distinguished from the squirrels by their pointed snouts. As their name would suggest, they are mainly arboreal animals, although they also spend time on the forest floor. They have quite good vision, a more complex brain than is typical for mammals of their size and usefully dextrous hands. Indeed, for animals that are not primates, they are remarkably primate-like.

For many years, the tree shrews were included in the primate order and it is only recently that they have been accorded one of their own. However, their similarity to the modern primates is superficial and lies in the possession of a relatively unspecialized skeleton and some adaptations that result from the similar demands of living in the same type of habitat.

The primate order is, of course, the second huge success story. Its precise origins are somewhat mysterious, but it would appear that by about 60 million years ago, small, insectivorous, clawed mammals had given rise to a number of forms that were more recognizably primates in terms of their increasing intelligence, forward-facing eyes with stereoscopic vision and grasping hands and feet that were equipped with nails rather than claws.

According to many biologists, the Plesiadapiformes were initially the most successful of the new order and they gave rise to a variety of species that specialized in eating the fruits and leaves of forest trees. They were, however, primitive by modern standards; and as time went on, they were eclipsed by the families of primates that we still see today. There is also an alternative view that forward-facing eyes with good stereoscopic vision – and hence good depth perception – and grasping hands and feet with nails rather than claws can be accounted for if the ancestral primate stock consisted of small animals that lived in a milieu of fine branches and fed on live insect prey. They would have located their victims by sight, crept carefully up on them and grabbed them by hand while hanging on with their feet. If this latter theory is correct, then it seems unlikely that the vegetarian Plesiadapiformes were actually primates. This question has yet to be resolved.

The modern primates

There are 51 genera of living primates arranged in 14 families. Together they make up the most widespread of any order of plants or animals. If brief migrations are included, they can be said to be found in every part of this planet, in almost every conceivable environment from the ocean floor to the outer atmosphere. In addition, they have been recorded in space and at a few locations on the Moon.

A large part of this distribution is accounted for by *Homo sapiens*; almost all of the remaining species are confined to the tropical and subtropical forests and savannahs of Africa, Asia and America. Non-human primates do not occur naturally in either Europe or Australasia.

Although the primates are a mammalian order with a taxonomic status comparable, for example, to the bats or the rodents, they are not easily distinguished by any particular characteristic. Whereas the bats are obviously adapted for flying and the rodents for gnawing, the primates lack a single, universal specialization, unless it be the ability to climb by grasping things with their hands and feet.

Their arboreal way of life led to the development of a complex of physical characteristics that tend to become more accentuated as the order is traced from lemur to human. The original mammalian pattern of five digits on the end of each limb is retained by almost all species, but flat nails replace claws in most cases, enhancing both manual dexterity and sensitivity. More efficient stereoscopic vision improves depth perception, thus making primates less likely to miss when they grab at branches or insect prey; and the all-important eyes are protected by tough, bony orbits. The combination of stereoscopic vision, delicate manual dexterity and a powerful grip

Distribution of modern non-human primates.

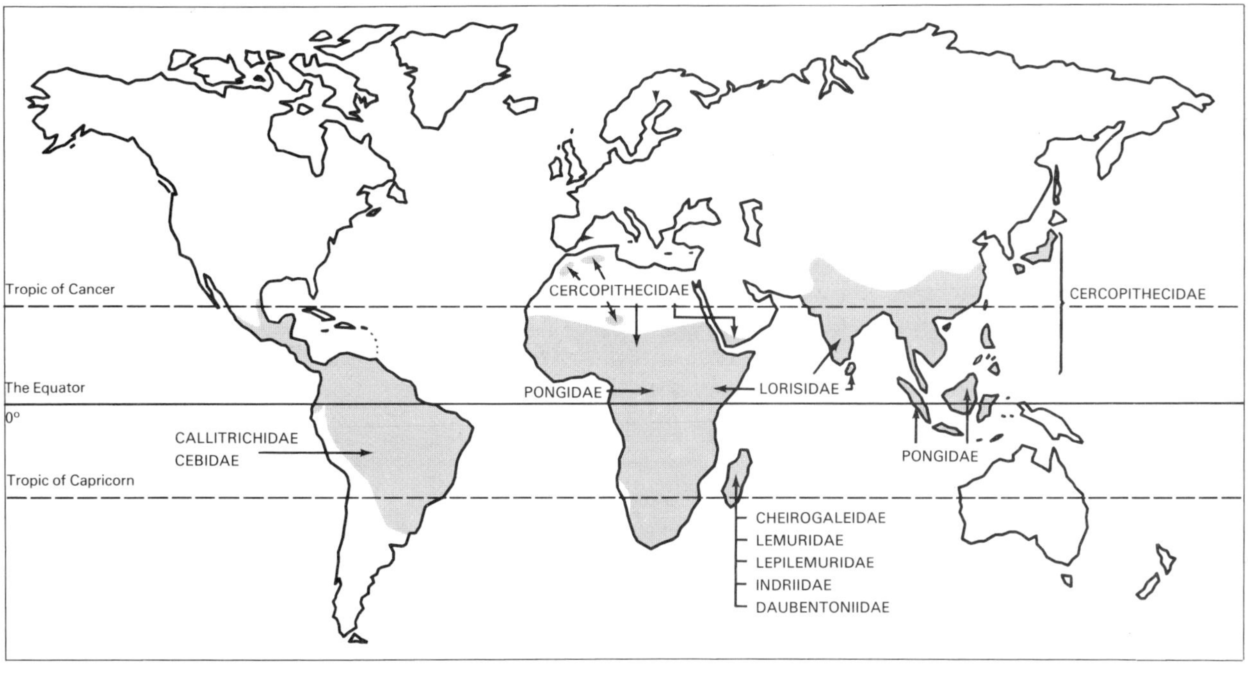

opened up new horizons in terms of plucking, manipulating, dismembering and so on. In humans, and to a lesser extent in the apes, it has also led to building.

As vision became more important, so smell became less so. This is reflected in the short snouts of most diurnal primates, in the reduction of the parts of the brain that interpret smells and in the loss of the rhinarium (the moist skin on the snout) by monkeys, apes and humans.

The back teeth, or molars, have remained relatively unspecialized, reflecting the rather varied diet that is to be found in the trees; and generally speaking, the arboreal life of the primates has given rise to various degrees of erectness of the body posture, including upright, bipedal hominids – even though we now seldom climb trees once we reach adulthood.

One of the most striking of the primate trends is that towards more intense and lengthy pampering of the young from the moment of conception. During pregnancy, the foetus is nourished by a particularly intimate mingling of its blood system with that of its mother. At birth, the infant embarks upon what will be a long period of growth and development in the care not only of its mother, but also often in that of other relatives or members of the same social group. The complex and relatively stable societies of primates provide, with some dramatic exceptions, a degree of protection that enables the young to have a long period of learning.

This educational period is crucial for animals that specialize in sophisticated reactions to immediate circumstances rather than in stereotyped patterns of behaviour. It is particularly useful when new environments present themselves and explains why some monkeys become efficient agricultural pests or burglars when human settlement encroaches on their natural habitat. It also explains why primates make thoroughly dangerous pets, since sooner or later, it invariably occurs to the nice, placid pet that it might be worth experimenting with a violent attack on a human being, preferably on the weakest or smallest one around.

The formidable intelligence of the primates probably got its evolutionary start as a result of the mental effort that was necessary for good eye–hand coordination. Since then, it has proved to be adaptive for all sorts of dealings with the ecological environment and has received a further spur to its evolution as a tool to be used in social competition. Humans are not the only primates for whom quick thinking, bluff and political manoeuvring can be a help to the ambitious. As primate societies became more complex, so there appears to have been no end to the usefulness of more and more intelligence, and in the twentieth century, the most successful primate has accelerated the process by using his biological brain to construct artificial ones.

When looking at the huge variety of modern primates, it is sometimes difficult to see the wood for the trees. It may therefore be helpful to think in terms of a few broad generalizations, bearing in mind that there are plenty of specific exceptions.

Very roughly, the order can be divided into the prosimians of the Old World, most of which are small and nocturnal, and the simians of both hemispheres, most of which are larger and diurnal. Within the simian grouping, there is a division into New and Old World forms that is perhaps of equal importance to that between them and the prosimians. The New World simians are all monkeys, and most of them are fairly small, and almost exclusively tree dwelling. The Old World simians include monkeys, apes and humans, and they tend to be bigger and more likely to spend time on the ground.

Most primates are vegetarian, either wholly or in part, with insects and spiders being more important in the diets of the smaller ones. Quite a few species top up their diets with meat.

Within any single species, the interests of the females do not coincide exactly with those of the males; and the forms of most social systems are ultimately dictated by the ecology of the females. The males may well be more spectacular, bigger, bossier, easier to observe, or just more interesting to the male observers who at one time formed the overwhelming majority of field primatologists, but that does not necessarily make them the arbiters of whether the species will be solitary, monogamous, polygamous, or promiscuous.

Since the successful production of large numbers of descendants is the characteristic that has been selected and refined by millions of years of evolution, most males do best by being promiscuous. If the females would only let them get away with it, paternity without responsibility and mating with as many females as possible would be all that would matter.

The females, however, do not as a rule let the males escape responsibility. They are not in a position to increase the number of their descendants by mating with large numbers of males because of the time and effort that are required for pregnancy and lactation. The strategy of a female, therefore, tends to be aimed at the successful rearing of however many offspring she can reasonably expect to produce in a lifetime of maternal effort. More than anything, this means that females need adequate food supplies and protection from predation. At the most basic level, they do not have to worry about finding the males; in one way or another the latter will be ready to make the necessary efforts.

Sometimes, the females of a species divide up the available land into territories that they defend against each other, thus ensuring their food supplies. When this happens, the males tend to divide up the land into territories of their own that overlap with those of as many females as possible, and therefore give them access to the maximum number of females. The males thereby compete for the females. This type of social system happens among many of the prosimians.

Monogamy is most likely to be a special case of overlapping territories, where a male can only manage to overlap with a single female, with whom he forms a special relationship. Invariably, the special relationship has evolved to a point of considerable complexity and sophistication. This type of social system happens among many of the smaller New World monkeys and among the lesser apes of Asia.

When the females collaborate with each other by living in a group, their collaboration tends to be lifelong and to provide the stability in the society, whether or not they defend a group territory. Many groups do not defend territories, and even single species are not always consistent about this: some modify their territorial behaviour to suit the local conditions. Where groups of females exist, males tend to attach themselves but in a less permanent way than the females. They hardly ever spend their adult lives in the group in which they were born, and they may move from one group to another of their own volition, or may be forced out of their groups by other males.

Under some circumstances a male can keep others out of the group, and a harem results; under other circumstances, more than one male is present with the females. Both of these types of societies are commonly found among monkey species. What they have in common with the overlapping-territory type is that the females gain some sort of security for their nurseries, while the males compete for sexual access to them. In the process, the males often end up protecting 'their' females in some way and being able to discriminate their own offspring (which they favour) from others (which they may even attack).

The final twist comes in those species where the males are unable to prevent the females from choosing among them for a mate. This allows the females to discriminate on the basis of different qualities in the males, and it is easy to see how this could be important to a female if the male is likely to be around during the rearing process. In a monogamous species, for example, the female would usually be best off to select a faithful male who will help her to defend an adequate territory – if there is some way in which she can do this. Female choice of a male could also be important in cases where he will be long gone by the time she has her offspring: she might choose strength or intelligence, for example, and so maximize her chances of having strong or intelligent offspring. A female will only be unable to choose if a male can rape her (and even then she may have a trick up her sleeve, as in the case of the orang-utan), or if he can keep all others away from her. In the latter case, he will, by definition, be good enough at his job to be an attractive proposition in that respect at least.

In summary, there are exceptions but females tend to be the more stable elements in primate societies, especially once they start collaborating with each other. When they are in a position to choose, females tend to go for male quality while males go for female quantity. Of course, neither works out the logic of their situation and tries to evolve in a certain manner, although that is a classic way in which behavioural evolution is misunderstood. What happens is that the individuals who happen to do the correct thing are more successful than those that do not, so 'doing the right thing' becomes more and more widespread in the population just as surely as does having the right physical characteristic.

The descriptions of the genera

In the pages that follow, each primate genus is illustrated by examples of some of the more typical, and some of the more interesting or surprising species.

When reading these descriptions, it is important to bear in mind that we do not by any means know all that there is to know about any primate species. Many of them have never been studied in the wild, some have never been seen in captivity, and a few are known only from one or two specimens that have been collected. There is even good reason to believe that a few new species are waiting to be discovered.

The descriptions are broad introductions to the genera; and the distribution maps are indications of where wild populations are believed to exist, given the availability of suitable natural habitat.

OVERLEAF **The thick-tailed bushbaby is the biggest member of its genus (see page 62).**

2. THE PROSIMIANS: THE 'ALMOST MONKEYS'

The biggest division in the primate order lies between the simians, the familiar monkeys, apes and people, and the less familiar prosimians. The latter group is comprised of a diverse array of different kinds of lemurs on Madagascar and some relatively small, nocturnal creatures from Africa and Asia.

All of the prosimians are united by the retention of certain very ancient mammalian characteristics upon which each species has evolved its own specializations. When viewing these animals as examples of primates, they can be seen, as a group, to lack many of the characteristic that we regard as typifying our order. This is best understood if we try to see why they are so conservative, on the assumption that such conservatism has been beneficial to them.

The answer can be obtained relatively simply by realizing that both the earliest primates and most of the modern prosimians were, and are, nocturnal.

The nocturnal ancestry of all but one modern prosimian species (including the diurnal ones) is given away by their possession of a 'tapetum lucidum' – a crystalline shield that lies behind the eye's light-sensitive retina. By reflecting light back onto the retina, the tapetum improves vision under poor light conditions. The presence of a tapetum in 16 prosimian genera is proof of its presence in some common ancestor which must have been nocturnal, or else why would it have evolved? It is virtually impossible that the same structure evolved separately many times; and even if some sort of reflecting shield had evolved on separate occasions in different prosimian lineages, there would be evidence of structural differences in the tapeta of different ancestries. This is not so, and it is much simpler to think of its inheritance from a common ancestor. In addition, the diurnal lemurs have tapeta that they do not need, and which must therefore be left over from a nocturnal past. The only prosimians that lack a tapetum are the tarsiers, but they are different in many ways and will be dealt with separately.

Being nocturnal, the earliest primates must have relied more on the senses of smell and hearing than on vision, thus combating the limitations of seeing in the dark. The use of acute hearing has been retained by most, or perhaps all, primates, but it is only the prosimians that have retained a primary reliance on smell. They do this both by their capacity to detect and interpret smells with a high degree of sophistication, and by their ability to leave 'smell messages' of various kinds for each other.

One important consequence of this is that all of them (except the tarsiers) have retained their ancestor's wet nose, something that is related to having an acute sense of smell. It is relatively fixed in place and goes with an upper lip that cannot be moved around very much. The simians have lost this arrangement in favour of a dry nose and a mobile upper lip, which permits very much more social communication by means of facial expressions – something that is only useful in daylight.

Thus the prosimians' 'primitive', rather expressionless faces and great reliance on the sense of smell are very understandable as the retention of features that are useful in the modern world.

Similarly, most nocturnal mammals are rather small, so the generally diminutive nature of modern prosimians is best regarded as being most appropriate for them. Tiny prosimians are not tiny because they got stuck at that small size millions of years ago; they are tiny because bigger individuals of their species are selected against by being unsuitable for their way of life.

Having given rise to the monkey and ape line long, long ago, the prosimians can hardly be expected to do so again, not least because the simians have now had millions of years to perfect the business of being diurnal primates and have taken up all the obviously available niches.

So the prosimians go on being small nocturnal primates. In fact, it may be because they are so good at it that the diurnal simians have not produced a viable nocturnal species, except in South America where there are no competing prosimians.

Equally, the only place where the prosimians have gone diurnal is on the huge island of Madagascar off the coast of East Africa, where there are no competing monkeys or apes. As might be expected, the diurnal species of prosimians show evidence of moving in a monkeylike direction by having become bigger and by living in cohesive groups as is typical of monkeys.

However, it is not at all surprising that the diurnal Malagasy lemurs have not dashed along an evolutionary pathway in the direction of becoming just like monkeys and apes, because evolution does not progress towards some ultimate goal. Rather, competition leads to the selection of the best that happens to be available and the fiercer the competition, the faster evolution will progress. In the case of Madagascar, its isolation has restricted the number of species there, which in turn has led to different kinds of competition from those found on the more species-rich continental land masses. Thus the modern species of diurnal lemurs have evolved in parallel to the simians only as far as the kinds of competition on Madagascar have made necessary in order for them to continue to be successful.

There is one other prosimian characteristic that should be mentioned as setting them apart from other primates, and that is the 'dental comb' or 'tooth scraper'. This is particularly exaggerated in some species and is made up of the front teeth of the lower jaw which project forward horizontally and in parallel, like a comb. Prosimians even have their own toothbrush to go with it, in the form of a horny structure under the tongue, known as the 'sublingula'. This is used for cleaning debris off the dental comb.

Among the prosimians, the front teeth of the lower jaw project forward to form a 'dental comb' that is used for scraping gum off trees and for combing tangles out of their hair. This is the lower jaw of a bushbaby.

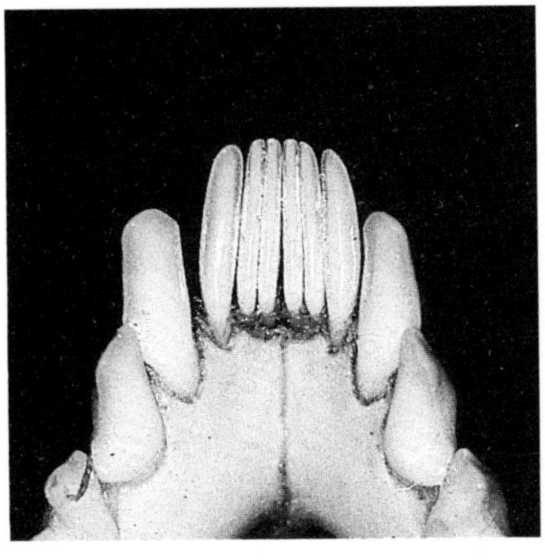

Prosimians use their projecting front teeth for scraping gum off trees and for combing the tangles out of each other's hair (and their own). This has given biologists the opportunity to have a fine controversy about whether gum eating or grooming was more important in the beginning, but we will pass that by. The important point is that with the exception of the tarsiers and one generally aberrant species, all prosimians have dental combs. In contrast, simian primates do not have dental combs.

I The Lemurs of Madagascar: five families and an array of species

Madagascar is one of the biggest islands in the world, being about 1600 km (1000 miles) long, nearly 600,000 km^2 (232,000 square miles) in area, and straddling the Tropic of Capricorn. There is a high central plateau, now largely devoid of natural, primate-supporting vegetation; but in the lower-lying country around this, there is a great variety of vegetational types. These range from moist rain forest in the east, through quite distinct, drier and more seasonal forest in the west, to severely arid country in the extreme south and southwest.

The ancestors of today's lemurs were isolated on this mini-continent about 50 million years ago, and since then they have diversified and proliferated in the absence of any monkeys or apes; and also in the absence of many other of the more successful mammalian species that are to be found just across the Mozambique Channel in Africa. Most importantly, they were isolated from most of the predators that are found elsewhere in the tropics.

About 2000 years ago, there were many more primate species on Madagascar than is the case today. Sadly, the biggest and some of the most spectacular lemurs have been wiped out by human activity, a process that gives every appearance of continuing today.

The species described in this section are what is left of the finest flowering of the prosimian line.

The mouse lemurs GENUS *Microcebus*

Imagine yourself in the rain forest of Madagascar. It is night and dense tangles of vegetation loom in the blackness on either side of the trail on which you stand. Something scuffles among the dead and rotting leaves at your feet. Suppressing the urge to flee the venomous snake or giant spider of your imagination, you take a closer look. It hops; maybe it's a frog? But it is grey and furry, with

big ears and a long tail – what kind of strange little beast is this?

If this ever happens, you will have found a grey mouse lemur, one of three species of the smallest of all primate genera. Adult grey mouse lemurs weigh between 40 and 100g (1.5 and 3.5oz); about 60g (2oz) is the norm. With their tiny size and their cryptic, nocturnal exploitation of leaf litter and dense tangles of small branches, they are surely the closest living approximation of our earliest, truly primate ancestor.

Conservative they may be, but mouse lemurs have been quite successful in their unobtrusive way. They seem to be able to utilize the small branches and finest vines of their forest habitat, whether these occur near to the ground or way up in the canopy at the tops of the trees. But usually, this type of vegetation is to be found furiously competing for the sunlight in openings caused by tree falls, or along paths, and that is where the mouse lemurs are most likely to occur. They are a classic 'edge species'.

They eat almost anything: fruits, flowers, leaves, sap, insects, spiders, tree frogs, chameleons, eggs and even small birds. Coquerel's mouse lemur, a slightly larger species from the drier forest in the west of Madagascar, even milks a certain kind of bug for a sweet exudate in a manner that is reminiscent of ants with their aphids. It must be the only primate that often starts its working night by going to lick its insects.

An even more surprising characteristic of the mouse lemurs is not, however, unique to their genus. As among the dwarf lemurs

The grey mouse lemur is one of the tiniest of primates. It finds food and safety in dense tangles of fine branches.

and the bushbabies, females regain their 'virginity' following each mating period. Their vaginas only open up for giving birth and for about nine days while they are in oestrus (heat). For the rest of the time, the organ is covered by furry skin. The growth process must be extremely rapid because there is evidence that at least some mouse lemurs manage to produce two litters during each annual birth season. There are two or three babies per litter.

The young are born and grow rapidly in nests that are sometimes shared by several females, all of which reproduce at about the same time. The grey mouse lemur nests in tree hollows or fashions a ball out of leaves, but Coquerel's species builds a bigger structure out of branches, vines and leaves that looks rather like a squirrel's drey. The infants begin their first forays out of the nest when only about three weeks old and are fully adult after a year.

All age–sex classes sleep in nests, singly or communally, and it is probable that any individual has several nests that it regularly uses and to which it can rush in the case of an emergency. Although mouse lemurs generally seem to forage alone, their societies are undoubtedly complex and stable social relationships are established between individuals. Scent-marking, either by urinating or by rubbing against objects, is an important means of communication, which brings us to another oddity about this genus. Coquerel's mouse lemur emits a strong odour every now and then as it goes about its business, and it seems that in doing so it is able to give its fellows such information as who it is, what sex it is, whether or not it is in oestrus and even how long ago it made the smell.

The dwarf lemurs GENUS *Cheirogaleus*

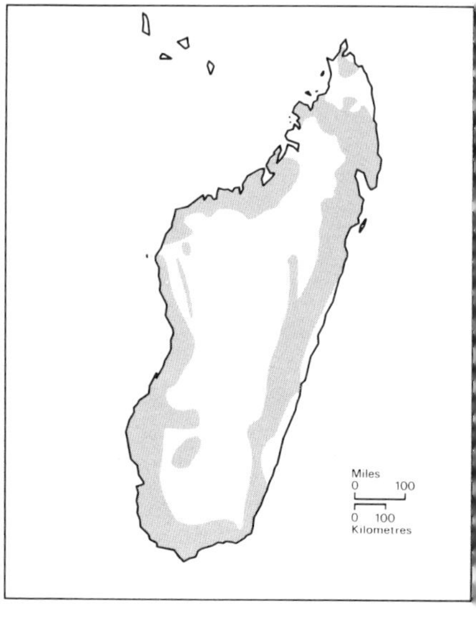

When the winter cold clamps down on the northern hemisphere, many of the resident animals go to sleep. Some sleep for a few days at a time and generally spend the winter feeding occasionally and doing as little as possible, while others truly hibernate until the spring. Their body temperatures drop almost to that of the hole or cave in which they are curled up, and they successfully avoid the need to eat at a time when food is hard to find. In the deciduous dry forests of western Madagascar, a close relative of the mouse lemurs, the fat-tailed dwarf lemur, uses a similar tactic to pass the difficult days of the southern winter.

There are three species of dwarf lemur, a rare and mysterious hairy-eared one, a bigger one in the wetter, less seasonal rain forests of eastern Madagascar and a smaller one in the west. Not much is known about the natural life of the first two species, but the third, the fat-tailed dwarf lemur, is unique among primates for being a true hibernator, even though it lives in the tropics and never has to combat freezing weather. It does, however, have to cope with an annual period of drought during which very little food is available; and this is what it avoids by hibernating. So efficient is it at switching off from the world that it can remain in its winter retreat in apparent suspended animation for eight months, if necessary. It typically does this in deep holes in tree trunks, where several individuals will spend the winter drought piled on top of one another and mixed up with their version of bedding – some nice, comfortable rotting wood. As in northern mammals, their body temperatures drop to a point just above that of their surroundings.

Although inactive and therefore in little need of fuel for their bodies, fat-tailed dwarf lemurs would starve but for the fact that they build up reserves of fat under their skin and in their tails – hence their name. During their active period, their body weights increase by over half, usually to about 220 g (7.7 oz), and the volumes of their tails nearly treble. Thus they eat until they become exceedingly fat, after which they go on the most extreme of crash diets, later emerging sylphlike again.

Mating takes place soon after they emerge from hibernation and the babies are born in January when plenty of food is available. Dwarf lemurs depend a great deal on fruit, but they are distinctly opportunistic and will also eat flowers, nectar, gum and the occasional chameleon. Insects, especially beetles, are particularly important at certain times of year. They are completely nocturnal and appear to do most of their feeding during the latter part of the night, returning to their nest holes by dawn.

Having assured their contribution to the next generation, the adults retire into hibernation as the dry season approaches, leaving their young to spend another few weeks in building up their fat reserves before they too retire for what will therefore be a slightly shorter period. It seems that full growth and development is necessary for a full-length winter sleep.

The dwarf lemurs are distinguished by another remarkable characteristic. They have the habit of leaving long, smelly smears on branches by dragging their backsides against the bark as they defecate: they have the peculiar ability to protrude their anuses while doing this; and they sometimes add urine as they go.

Scent-marking is common enough among mammals (witness any dog marking lampposts and trees), but the dwarf lemurs' efforts must be unequalled for durability. The likely explanation for this is that it may survive the winter drought and so establish and maintain an individual's claim to reside in a particular area, even though that individual may still be asleep. To late risers in the spring, this might be important and save argument about who lives where. Some support for this suggestion comes from the observation

OPPOSITE **Flowers are among the most important items in the diet of the fat-tailed dwarf lemur, but it does not usually emerge from its nest-hole during the day.**

that fat-tailed dwarf lemurs do most marking in the trees as the winter drought approaches. Presumably, the benefits are similar for greater dwarf lemurs which make similar marks, although the whole effect must be speeded up: the marks will wash away quicker in the wetter climate of their rain forest, but the markers will not be asleep for as long as their western cousins.

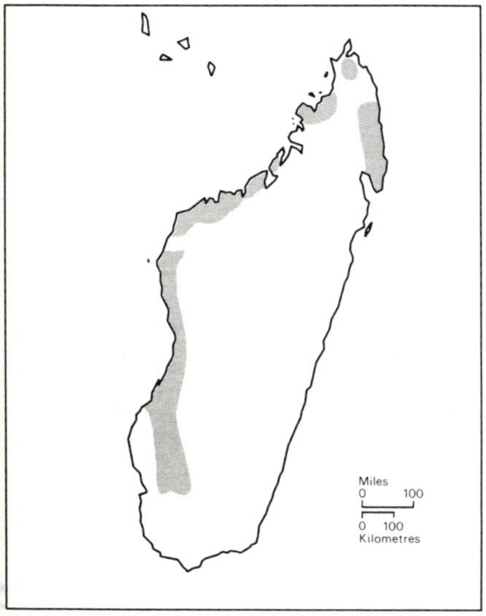

The fork-marked dwarf lemur GENUS *Phaner*

It seems very likely that the mark that gives this animal its cumbersome English name is also a convenient label for the creatures themselves. Active only at night, they are unlikely to obtain a very good view of each other when they meet, so it must be convenient to be able to note that the squirrel-like creature scrabbling down the tree trunk nearby wears the species badge: a dark spinal stripe that splits between the ears and continues as two definite stripes to the eyes. It is not unusual for nocturnal mammals to have some definite contrasting pattern that will be mutually displayed when two individuals peer at each other through the murk.

Of course, what happens then depends on how the two individuals feel about each other. In the case of fork-marked dwarf lemurs, relations are unlikely to be too friendly unless the two are of opposite sexes. As far as is known, fork-marked dwarf lemurs are usually monogamous, although bachelors and male bigamists do

RIGHT **The characteristic stripes on the head and back of a fork-marked dwarf lemur signal its membership of the species to other individuals.**

occur. The pairs live in small territories in which they share a nest hole they have quite possibly inherited from a mouse lemur.

They also feed together, or at least the male follows the female around as they move from tree to tree. They keep in constant touch by calling softly to one another. Indeed, they are quite vocal creatures and they are distinctly noisy when excited or alarmed. Neighbours often spend time calling to each other across their territorial boundaries, and males make their presence felt with a loud call that they alone can emit. As with many birds in an English woodland, land ownership is advertised by calling.

Intolerant of their neighbours though they may be, relations within the family are close, and it is quite common for two fork-marked dwarf lemurs to hang upside-down with their feet from a branch while they groom each other with their hands. Grooming also takes place in less hazardous situations and the males have large skin glands on their throats which they rub affectionately over their partners. Presumably, the male likes to sniff his scent on the female.

The bond between the pair affects most aspects of their lives and certainly is not limited to a sexual attraction. Like mouse lemurs, the female only opens up in the breeding season and in this species, she may only make herself available for two or three days each year. Little is known about what happens after that, but apparently the young are suckled and weaned when the weather is wettest and most food is available in the forest.

Fork-marked dwarf lemurs specialize in the collection of gum. Their favourite way of starting the night is to scrape tiny drops of gummy exudate off trees where the bark has been penetrated by a beetle larva. In addition, the lower front teeth are long and forward-pointing in an extreme version of the lemuroid 'dental comb'. Together with rather pointed upper teeth and a long tongue, the dental comb is used to scoop gum out of wood that has been riddled with tunnels by insect larvae. To do this, the animal has to be able to move freely over the trees' surfaces, so it has very sharp, long nails that are ideal for climbing on large branches and slippery trunks. Being about 300 g (10.5 oz) in weight, these small primates really are very reminiscent of large squirrels as they scurry up and down the trunks on rather fixed itineraries of visits to good gum trees.

There is only one species of fork-marked dwarf lemur and the secret of its success appears to lie in its ability to make a living out of eating mostly gum and sap, with insects being taken for their protein. It is an extreme specialization that sets it apart from the other Malagasy primates, but one that has been seen to occur more than once among the bushbabies of Africa and some South American monkeys.

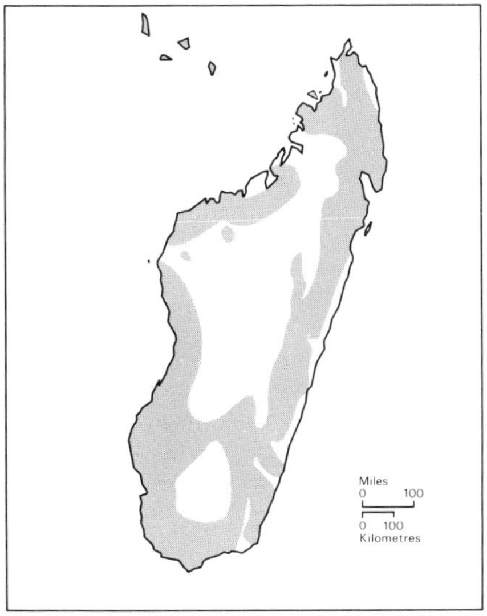

The true lemurs GENUS *Lemur*

This is the most widespread and diverse of the Malagasy genera, with six species showing a variety of solutions to the problems posed by different ways of life. The true lemurs come closer than any other prosimian genus to displaying the range of ecological opportunism that is displayed by the two great Old World monkey genera, the guenons and the macaques.

The best-known species is a little bigger than a domestic cat, with rather long hind legs and short forelimbs, which together give its members a pronounced bottoms-up look when they run around on all fours, as they commonly do. Their tails appear to have been borrowed from raccoons – hence the name 'ring-tailed lemur' – and their foxy faces have black, panda-like eye patches that give them the air of masked highwaymen. Individuals weigh about 2 kg (4.5 lb).

Groups composed of one or two dozen ring-tailed lemurs may be found in the relatively dry forests of southern Madagascar, where they are most likely to be noticed either because of the noise they make or because of the conspicuous nature of their tails. As lemurs go, they could hardly be described as shy and retiring. They start their day at about dawn when, after some preliminary movements around the branches of the tree in which they have been sleeping, they begin looking for their breakfast or, if the weather is suitable, they do some sunbathing spreadeagled among the branches.

Ring-tailed lemurs are vegetarian in their diet, preferring fruit but also taking leaves and flowers. A group will usually pass the day by concentrating on a few good feeding areas, often eating first in the trees and then descending to the ground to browse or to pick up fallen fruit. They forage and travel at all levels of the forest, commonly running along branches or leaping between them, landing in such a way that their powerful hind legs touch first. Although they spend about 15 per cent of their time on the ground, they are never far from the safety of the trees and will leap up into them at the slightest sign of an alarm.

There are usually more adult females than males in the groups and they are dominant over the males. They are probably more important to the organization of their society, since females remain all of their lives in the group in which they are born, and thus have the opportunity to form strong, stable relationships with each other. Males, on the other hand, switch groups at least once, and possibly several times, in their lives. However well a male is established, he is certain at some point to get the urge to leave. This is in spite of the fact that the resident males in the new group of his choice are very likely to make strenuous attempts to drive him off. At certain times of year, particularly following the birth season, the forest may

Ring-tailed lemurs not only show off their tails in a constant visual signal, they also coat them with scent from glands in their skin and then shake them at each other in 'stink-fights'.

have many single males or small all-male parties wandering about as they seek out new groups.

This transferring is hardly surprising since it is almost universal among mammals and birds to have some way of avoiding incest; and in group-living primates, this end is achieved if one sex remains while the other wanders off.

Each group lives in a small home range that contains all the necessities of life and usually overlaps slightly with neighbours' ranges. The part that does not overlap is defended as the group's sacred territory, with permanent members – the females – doing most of the defending. Territorial encounters tend to consist of opposing groups of females running at each other and uttering threatening calls until one group retreats.

Males do fight, but mostly with each other. Although some aggressive encounters may consist of as little as a threatening stare, others escalate through lunging and cuffing to hair-pulling and serious biting. Males also indulge in 'stink-fights', a unique form of warfare that makes use of the ring-tailed lemur's two main external glands. These are situated on the upper chest and inner forearm. Both exude strongly scented secretions and the latter also has a black spur along with the glandular part.

In a stink-fight, a male briefly rubs the two glands together then stands up on his hind legs with his tail held forward between them. He then pulls the tail down between his forearms, repeating the movement several times. His spurs are pulled right through his tail fur, deeply impregnating it with the scent from the glands. All the

The baby is a near-perfect miniature of the mother and will be totally independent in well under a year.

while, he stares at his opponent with a characteristic, rather dog-like expression and makes squealing or purring noises. He then drops onto all fours and shakes and quivers his tail over his head at the object of his display, who receives a great waft of offensive smell. There is quite a difference between the sexes in how these stink-fight displays are received. Males usually run away whereas females are more likely to attack their protagonists.

As in many animals, signs of aggression reach their peak during the mating season when a forest with many ring-tailed lemurs may present a scene of mild pandemonium as the animals mate, fight, howl and fall out of trees. However, the life of a ring-tailed lemur is by no means all aggression and excitement. The females spend much of their time caring gently for their infants, the young play with each other, and members of all age groups express the more affectionate side of their natures by nose-touching and grooming each other. In addition, much of their time is spent quietly dozing, especially when the weather is at its hottest.

The other species of true lemur may not be quite as flamboyant as their ring-tailed cousin, but they also exhibit some interesting idiosyncrasies. The smallest of them, the mongoose lemur, has the distinction of being found furthest from Madagascar of any lemur, on three of the Comoro Islands. Like all lemurs, however, it is also found on Madagascar itself. How it reached the Comoros is not known, but some human help is suspected.

It is a thickly furred, grey-brown animal, about 90 cm (3 ft) from the tip of its whiskered nose to the end of its tail. The tail accounts for about half of its length. It is something of a sweaty beast, having many glands in its skin, and scent-marking appears to be quite an important means of communication. It regularly marks the branches in its environment by rubbing them with its crotch, forehead or hands, all of which leave a trace of smell behind.

Like its ring-tailed relative, the mongoose lemur is vegetarian by nature, preferring fruit but able also to eat leaves and flowers. It appears to be thoroughly adaptable, able to change between diurnal and nocturnal routines as circumstance dictates. In the seasonal forests of northwestern Madagascar, it is active by day during the cooler wet season and concentrates on the fruit of its favourite trees. When the dry season comes along, it has a double incentive to switch shifts. Firstly, the days become too hot for it to remain active for long; and secondly, big kapok trees produce a rich food source – but only at night.

The flowers of these trees are pollinated by bats, and to attract the bats, they produce copious quantities of sweet nectar that is rich in both energy-giving sugars and the amino acids from which animal proteins are built up. Kapok trees even make access to their flowers easy by conveniently shedding all of their leaves during the flowering season. The tree benefits from the animals that visit its flowers because they pick up pollen on their fur and later rub it off on other flowers of the same or different trees, so pollinating them. The flowers themselves are probably poisonous when fully developed: at any rate they are not eaten by the nectar collectors.

As the kapok flowers are only open at night, the mongoose lemur gets a strong food reward for becoming nocturnal. The lemurs run quickly among the branches, poking their long snouts into flower after flower and licking the nectar from deep inside.

Unlike ring-tailed lemurs, mongoose lemurs live in small family groups that contain a single mated pair and their immature offspring. As soon as the young reach maturity, they strike out on their own – although whether they do this of their own volition or because their parents give them a push is not clear.

An exception to this rule occurs on Moheli Island, one of the Comoros, where bigger aggregations are found, strongly suggesting that mongoose lemurs have enough behavioural flexibility to break out of the family pattern once in a while. It even seems possible that this happens on a seasonal basis, presumably with families getting together at certain times of year and splitting up again later. If so, this would be a unique arrangement for a primate, even though it is a familiar one among some water birds that pair for life but flock together in the winter.

Mongoose lemurs are also a little birdlike in showing a slight form of 'sexual dichromatism': differently coloured sexes. The cheeks and beard of the male are reddish, whereas those of the female are white. Sexual dichromatism occurs in a few primate species, but it is generally rare among mammals. It is another true lemur, the black lemur, that shows one of the most extreme forms among the primates, in that both sexes are born black but the females turn red-brown when they are about six months old.

Black lemurs are very closely related to brown lemurs, yet another *Lemur* species. These animals are remarkable for having hit upon their own solution to the problem of whether to be active by day or by night: they manage both by the simple expedient of interspersing periods of activity with periods of rest throughout the 24 hours of the day.

When awake, they spend most of their time foraging in the canopy levels of the trees, where they run along branches quadrupedally but are not averse to leaping about, even between vertical supports such as tree trunks, to which they can easily cling. From the point of view of the nectar-eating mongoose lemurs with which they share some forests, their diet is unfortunate in that it includes the flowerbuds of the kapok tree. Every bud that is eaten is one fewer of the nectar-producing flowers for later on. This means that brown and mongoose lemurs are ecological competitors.

The brown lemurs have evolved a social organization that is extraordinary for prosimians, although it has developed completely separately among the New World monkeys and the great apes: this is the 'fission–fusion' society in which the community is large and mixed, and individuals travel around in groups that frequently change both their size and their component individuals. Spider monkeys and chimpanzees do much the same.

Interestingly, brown lemurs, mongoose lemurs, spider monkeys and chimpanzees are among the primates that exhibit a pattern of behaviour once thought to be rare among apes, and non-existent among prosimians and monkeys: voluntary food sharing. Few primates are at all generous when it comes to the matter of food, so it seems surprising that one brown lemur will sometimes hold a tomato-sized fruit of the houbouhoubou vine while another individual feeds upon it. Or a fruit may be split in half and shared by two adults sitting peacefully side by side. Of course, there are plenty of occasions when they refuse to share, but the fact that they do so at all is an appropriate reflection of their rather cosy, peaceable natures. They spend a lot of time huddled up together, or grooming each other while in a tangled embrace.

The food-sharing behaviour of the mongoose lemur has only been observed once, but it is even more surprising since the instance involved apparently very purposeful 'giving'. On the occasion in question, a male–female pair had found a fresh coconut that had recently been holed, probably by a rat. The female repeatedly dipped her hand through the hole into the coconut and then she and her mate took it in turns to lick the milk off her fingers. It seems that she

ABOVE LEFT **Among black lemurs, it is only the males that are black. Here two of them are seen with a single female.**

ABOVE RIGHT **The little-known crowned lemur is found only in a very restricted part of northern Madagascar. The striking eyes of this male are typical of the lemurs.**

usually had first lick, but then it was her fingers they were using.

There is a great deal more to be discovered about this remarkable genus. The red-bellied and the crowned species are little known; and many of the better-known species have quite divergent races in different parts of Madagascar. Such divergence typically takes the form of different coat patterns and colours, but given the variety of Malagasy forest types in which they are found, behavioural differences are also likely.

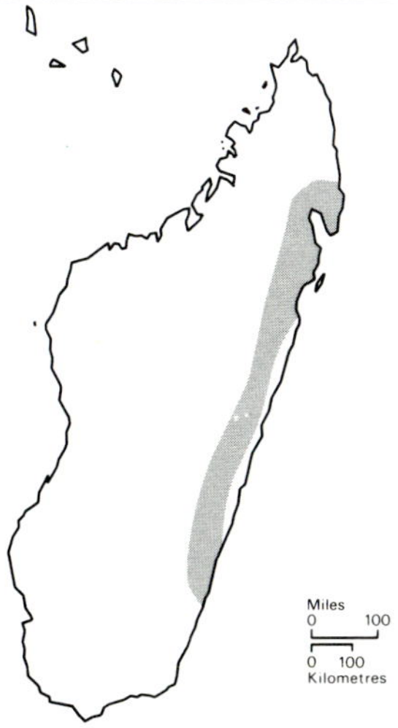

The ruffed lemur GENUS *Varecia*

In the eastern rain forests of Madagascar there lives a lemur that is not only one of the most beautiful of its family, it also has the most dramatic call. The raucous, in-and-out, sawing, bellowing barks of a group of ruffed lemurs, mingled with their occasional neighing noises, would be more appropriate for an impossible herd of horse–buffalo hybrids than for primates. At close quarters, it is deafening; and since the chorus often starts very suddenly, it would be an understatement to call it surprising. For an encore, ruffed lemurs follow the first crescendo with loud, rather avian clucks. To add to the effect, these vocal displays are most likely to be given as night descends and the darkness adds a forbidding dimension to the already mysterious and fascinating forest.

Ruffed lemurs start life by being left in the fork of a convenient tree or on some other suitable perch, rather than by being carried around. The female often gives birth to twins or triplets; and she

There is only one species of ruffed lemur, but there is much variation in patchwork colour-schemes and it is one of the noisiest of all primates.

makes them comfortable by pulling out her own fur to line their resting place.

Anatomically, these animals are something of a mixture between those that cling to vertical tree-trunks and leap between them, such as the sportive lemurs and sifakas, and the more quadrupedal forms. It seems possible that they have retained the ancestral body type that gave rise to the two extremes and as a result, they can do both quite well but are not as good at either form of locomotion as the more specialized genera.

There is only a single species of ruffed lemur, but it comes in at least four different coat colour and pattern varieties, all of which are based on the same patchwork theme. Whether or not the different colour schemes denote different races of ruffed lemur is a moot point: all are extremely impressive. There is a black-and-white variety in which a predominantly black face is surrounded by a brilliant white ruff over the ears, and long white fur on the forearms contrasts with black hands, upper arms, shoulders and upper torso. The effect is completed by a white lower torso and black-and-white legs. There are at least two other arrangements of black-and-white pelage, and an even more beautiful variety in which nearly all of the white is replaced with orange-red. This last form retains a white patch on the back of its neck which it dramatically displays when it hunches up into its submissive posture.

Ruffed lemurs have been known to science and admired since the eighteenth century, but not much has been discovered about their habits.

These lemurs are found living in small groups that are probably based on monogamous family units. As far as is known, these groups do not defend territories from each other but they do give loud calls to their neighbours, often in a sequence that is given by one group after another. Such 'calling rounds' are heard among many other primates, especially among the Old World monkeys.

The gentle lemurs GENUS *Hapalemur*

The reedbeds in the shallows of a fair-sized lake would hardly seem to be the most promising of places to look for a primate, yet there is a population of gentle lemurs that lives exclusively in such a place. The land around Lake Alaotra in eastern Madagascar has been completely deforested, but grey gentle lemurs may be found living among the papyrus or even swimming dog-paddle in the water itself. Sadly, their days may be numbered because they are preyed upon by the local fishermen and it is an easy matter to burn wide areas of reedbeds, driving whole groups of gentle lemurs out of shelter and into the nets and hands of waiting hunters.

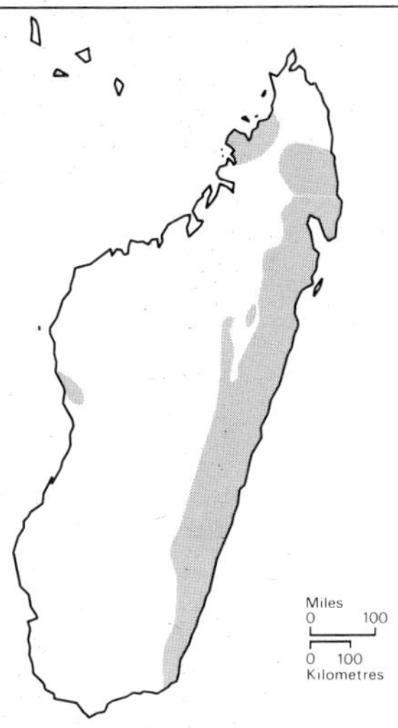

There are two species of these attractive, rather wet-nosed and furry-faced creatures. Both have thick, greyish fur, which often grades into reddish or sooty shades; and their tails account for half of their total length. The better-known species is the grey gentle lemur, of which there are three distinct races in different parts of Madagascar. They weigh a little over 2.5 kg (just under 6 lb).

The other species, the broad-nosed gentle lemur, was thought to be extinct until quite recently when a few were found in bamboo forest in Madagascar's wet eastern zone. Even so, it is extremely rare and very little is known about its habits. It is slightly bigger than the grey species, with which it has been seen to associate among the bamboo. The two types probably have much in common and it is known that both live in small groups that are likely to be families with a single pair of adults.

Their numbers seem to have declined markedly in recent years and not just in the reedbeds where they are rounded up with fire. They may also be diminishing in the forests in spite of the fact that their liking for bamboo should give them an advantage in areas where man's destruction of the natural forest has permitted bamboo to flourish. In such places, the gentle lemurs concentrate their diet on bamboo shoots, thoroughly chewing the soft parts out of small twigs before discarding the tough, woody bits.

Grey gentle lemurs are cryptic little beasts with a variety of soft and high-pitched calls by which they carry on a considerable, but discreet, amount of communication with each other. Such discretion is probably only too appropriate for such a small mammal; it may well be subject to the attentions of predators other than man.

Gentle lemurs may be similar to the ruffed lemurs in retaining some ancestral characteristics. Like the ruffed lemurs, the mothers do not carry their infants around all the time. When very young, the babies may be carried in the mother's mouth but, more often than not, they are left for long periods to cling to any convenient leafy branch. Later, they are carried on their mothers' backs, an appropriate arrangement for the semi-aquatic gentle lemurs of Lake Alaotra, since it is one that does not drown the youngsters when their mothers go for a swim.

Again like the ruffed lemurs, the gentle lemurs seem to be compromise generalists in their mode of locomotion. They run well on all fours, yet may also cling to vertical supports between which they can leap with agility.

Grey gentle lemurs have been variously described as nocturnal and diurnal; and the truth is possibly a combination of both, perhaps in a similar way to the brown lemur. Alternatively, they may turn out to be crepuscular, avoiding the dead of night and the heat of the day, and being active in the hours on either side of sunrise and sunset.

For illustration of a grey lemur (see page 44)

ABOVE LEFT **The grey gentle lemur is expert at clinging to the vertical tree-trunks of Madagascar's forests, where it avoids predators by being as unobtrusive as possible (see page 43).**

ABOVE RIGHT **Sportive lemurs are among the least active of primates, spending much of their time simply looking around from some convenient vantage point. Their name derives from the typical boxer's stance that they adopt when threatened.**

LEFT **The eyes of nocturnal prosimians are adapted to enhance light-and-dark contrasts by being most sensitive to blue-green, dimly lit objects. In the gloom of the forest at night, this helps them to stand out against the prevailing dark, slightly reddish background where most of the little light that is present is reflected off foliage. This is** *Lepilemur dorsalis*.

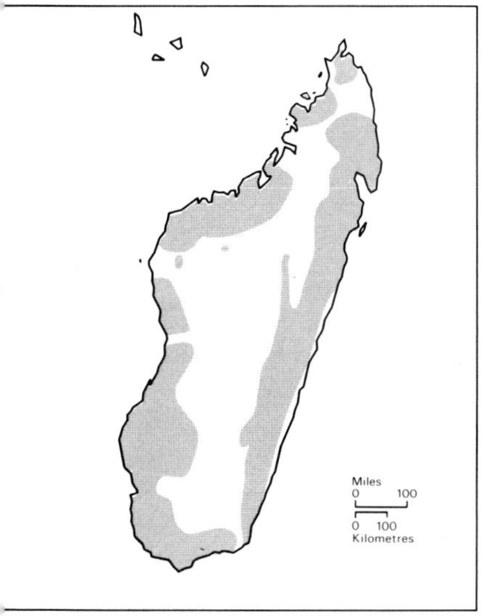

The sportive lemurs GENUS *Lepilemur*

The seven, very similar, species of this genus live a life that is dictated by a simple, if rather extreme, conservation philosophy. Sportive lemurs have found success on a low budget by following the principle that the less you spend and the more you save, the less you need.

Their diet consists of a large quantity of very tough foliage, eked out in the more arid parts of Madagascar with flowers for meal after meal whenever foliage is hard to come by. This is one of the poorest diets of any primate; more than three-quarters of it consists of cellulose and related compounds that present considerable problems for a mammalian digestive system.

In order to need as little food as possible, sportive lemurs do as little as possible. They are wholly nocturnal, passing the daytime in tree holes, except on the predator-free, small island of Nosy Bé where they simply roll up into balls on exposed branches. When night falls, they start their somewhat simplified routine of foraging and socializing. Foraging consists of finding a good food source and then feeding in separate browsing sessions of a few minutes each. More often than not, the lemurs sit immobile between browsing sessions, but once in a while they move to another tree by means of a series of rapid leaps. Such moves are seldom longer than about 20m (22yd), after which the animal settles down again to its feeding-or-immobile routine.

The moves are made with superb economy. In the first place, there is economy of direction: a leaping sportive lemur almost certainly knows exactly where it is going and what it will find when it arrives, because it lives in a very small territory with which it is doubtless exceedingly familiar. Ever budget-conscious, it defends just enough land to cater for all its needs and excludes other sportive lemurs of its own sex so that they cannot make use of its resources.

Secondly, there is economy of energy in movement. Sportive lemurs are most at home hanging onto large, vertical tree trunks or branches, between which they leap by pushing off with their powerful, long hindlimbs. Thus, their mode of locomotion calls for sudden, brief bursts of energy; and they have a trick for making an extra profit out of this. The moves that each makes during the night are spaced at fairly even intervals; and they are most likely to occur just at the time when an immobile sportive lemur has cooled down in the night air and some muscular exercise would be helpful for warming it up again.

Thus, instead of sweating or panting off excess body heat from activity, which is what most mammals do, sportive lemurs move when they need the heat as well as the new location.

Of course, their lives are not quite as perfectly ordered as this all of the time, and the scars that most adults have on their faces and tails bear witness to the competitive side of their nature. Since they defend territories, they must occasionally be put to the test by land-hungry rivals. However, in any established community of territorial animals, there is always a simple system by which territory owners may publicize that they are in residence as usual. Loud calls are commonly used among primates, and sportive lemurs do call at each other, especially on moonlit nights when rivals are most visible. But they also save on energy by a system of visual surveillance: they sit for hours where they can see and be seen. Males, more often than females, spend long periods of time just staring at each other across their mutual boundary. When one moves to a new location, the other typically does the same, and they settle down again to face each other.

Such visual advertising of land ownership is rare among primates and it is probably made possible in this case by the tiny size of the territories: 50m (55yd) by 50m would not be particularly small. Whereas animals of the same sex are excluded, the territories of males and females overlap. Those of the males tend to be slightly the larger and a successful male may defend an area that overlaps parts of the domains of two or three females. Nevertheless, relations between the sexes do not appear to be very close and adult individuals sleep separately.

Even though it conserves as much energy as possible, a sportive lemur's diet would not be adequate if it did not have another conservation trick in its repertoire. At intervals during its diurnal period of rest, it wakes up and begins licking its fur. Soon it concentrates on licking itself between the legs, sitting with its thighs spread wide apart and its tail curved forward to complete a sort of bodily bowl. Every now and then, it lifts its head and swallows some faeces that it has passed while licking itself.

Revolting as this sounds, it is a lifeline to the sportive lemur because it cannot digest all that tough, fibrous foliage without the assistance of millions of bacteria that live in its caecum. This is a blind alley in the lower part of the digestive tract at the end of which is found the appendix that we humans find such a nuisance. Because the caecum is so far down the alimentary canal, the nutritive products of bacterial breakdown cannot be digested unless they start again at the top end. Thus, the coprophagous (faeces-eating) sportive lemur is not ingesting just anything that passes, rather it is recyling material that has been made protein-rich during its first time around.

This behaviour is known to science as 'caecotrophy'. Rabbits do it as well, being likewise unable to survive without it on a leafy diet, but the sportive lemurs are unique among the primates in being caecotrophes.

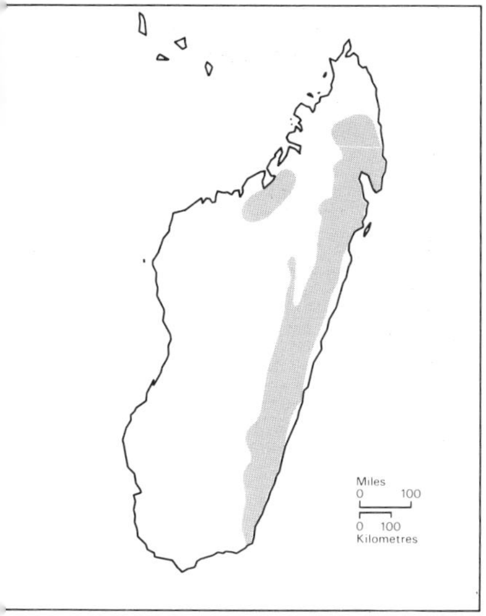

The avahi GENUS *Avahi*

Nocturnal and the very souls of discretion, the avahis have managed to hang on to most of the secrets of their existence. Sometimes known as woolly indrises, they are the smallest members of the indriid family, the others being the indris and the sifakas. An average avahi is only about 30 cm (1 ft) from the tip of its nose to the base of its tail, although the tail itself is rather longer than that. They are greyish-brown creatures with soft, thick fur and peaky faces that often wear a rather startled expression. They live in family groups, and by day sleeping balls of them may be found in dense foliage on the branches of trees.

Their large salivary glands, capacious stomachs and long caeca are all adaptations that would help with the digestion of a rather leafy diet, suggesting that much of their food intake is foliage, although this is likely to be supplemented with fruit whenever it is available. It is possible that avahis have the ability to subsist on leaves more as a fall-back tactic than as a matter of preference. There is usually more nutritive value in a fruit than in a leaf, but in the tropics fruit is more often hard to come by than leaves.

It is unlikely that avahis spend much time on the ground, but when they do, they progress in a series of kangaroo-like, bipedal hops. As with the other members of the indriid family, they are highly specialized for leaping from tree trunk to tree trunk, kicking off with both powerful hindlimbs and twisting in the air to land feet first. They are able to rest on vertical trunks by clinging on with hands and feet while they press their tails onto the tree for support.

Their method of progression seems to be somewhat hazardous for a creature that is active during the hours of darkness, but, presumably, they avoid disaster by a mixture of excellent night vision and a precise knowledge of their immediate surroundings.

Most of their calls are either very high-pitched or rather weak, making them difficult to locate by their voices. The avahi's alarm call is typical for a vulnerable bird or mammal because it is an unmistakable signal, yet it gives a predator little or no clue as to the caller's exact whereabouts. It is a brief, high-pitched note that is difficult to locate, partly because reflections off the foliage complicate the way in which the sound travels and because high notes are heard equally by the two ears of a predator. Lower-pitched and more complex calls are not 'bent' as much by the vegetation and they can sound different to the left and right ears. It is this difference that is used to determine the direction from which the sound came. So, when an avahi sees a predator and gives the alarm call, none of the animals that are within earshot is likely to know where it came from. This confuses the predator but it makes no difference to the other avahis, because they will all dive for cover anyway.

For illustration of an avahi (see page 48).

ABOVE With its enormously long legs, the avahi progresses from tree to tree by means of tremendous leaps (see page 47).

RIGHT In the arid lands of southern Madagascar, Verreaux's sifaka depends on the juicy flesh of succulent plants for its water supply. The plants belong to the family Didiereacea and are unique to Madagascar.

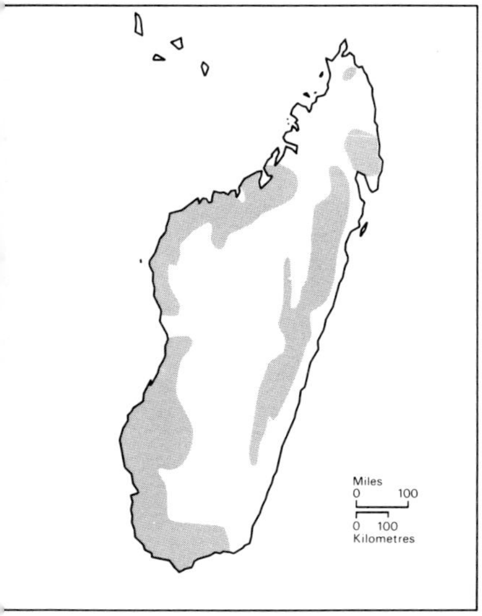

The sifakas GENUS *Propithecus*

In Madagascar's arid southern zone, cactus-like plants point their straight, multiple stems up at a clear, blue sky. Thick, succulent leaves grow directly from the trunks and thorns are everywhere. As in any semi-desert, there is beauty in this place, but only those skilled in survival may endure.

One such is Verreaux's sifaka (pronounced 'she-fa[r]k'), a dark-faced primate with a silky white coat, a chocolate cap and a long white tail. Together, its head and body are about 50 cm (19.5 in) long. It has relatively short arms, but the full extension of its enormously long legs would make it stand tall enough to bring its head level with a man's waist. In the wild, this is only obvious when it uses those legs to leap from one trunk to another and flies through the air in a perfect arc. As in the other Malagasy primates that biologists classify as 'vertical clingers and leapers', the legs are powerful springs for sudden take-offs and soft landings.

Four different races of Verreaux's sifaka inhabit the various forest types of southern and western Madagascar, and in the eastern rain forests there are five races of the slightly larger diademed sifaka. All are more or less the same in body type, but there is much variation in coat colour. Some have attractive orange patches on their fur; and dark, melanistic individuals crop up in the southernmost race of Verreaux's species.

Sifakas are diurnal, but wherever they are found they are like the nocturnal avahis in being able to survive on leaves for long periods, if necessary. In the dry season in the west, buds and mature leaves see them through until the rains bring out more tender young leaves, flowers and fruit. They also show a taste for eating bark.

An important factor in their survival in the arid south is their ability to eschew the drinking of fresh water. In such a place, the plants must store water in their flesh to survive, so the sifakas find enough liquid by choosing the juicy parts of the vegetation. For this to be sufficient, their body chemistry must be adapted something like that of a desert rodent, not to waste a drop. This is probably a major difference between the two species, since the diademed sifaka has no need of such refinements in its wetter, less seasonal habitat.

Verreaux's sifaka has a strange society for a primate. It lives in small groups, the composition of which appears to be stable, and within which the several males and females maintain a 'pecking order' or 'dominance hierarchy' that determines priority of access to food. But when the annual mating period comes along at the end of the wet season, the mature males spend as much time searching other groups for eligible females, and fighting for access to them, as they spend paying attention to those in their own group. The

The beautiful diademed sifaka inhabits the moist rainforests of eastern Madagascar, but little is known about its ecology or behaviour.

precise reasons for this behaviour are probably quite complex and connected with the advantages of mating with unrelated individuals, but it certainly makes for a highly competitive arrangement among the males in which some must be big winners and others equally big losers in terms of their success at fathering offspring. No such competition exists among the females, of course, although they may have an interest in selecting the best and toughest males to father their babies.

The young are born singly in the middle of the dry season and are carried on their mothers' bellies for about three months. After that, they switch to riding on their mothers' backs for three or four months; and then they have to do their own leaping.

They grow up in a pleasant community in which friendship between individuals is expressed mainly by sitting together, grooming and playing. There is no doubt that sifakas are intelligent enough to form strong personal likes and dislikes within their groups, and those that do form mutual attractions spend peaceful interludes scraping their dental combs through each other's fur and assiduously licking their partners. Sometimes they get so involved in competing to groom the best bits of each other – to a sifaka, this means the genitals – that they end up playfully wrestling rather than grooming. Both grooming and play are common among primates, but the latter is rare between adults in most species. Among sifakas, the adults are as likely to play as the juveniles and it is not unusual to see two full-grown males doing so.

Sifakas are surely among the most delightful of primates and by living successfully in so many different types of forest, from the wettest to the driest on their huge island, they show an impressive degree of adaptability.

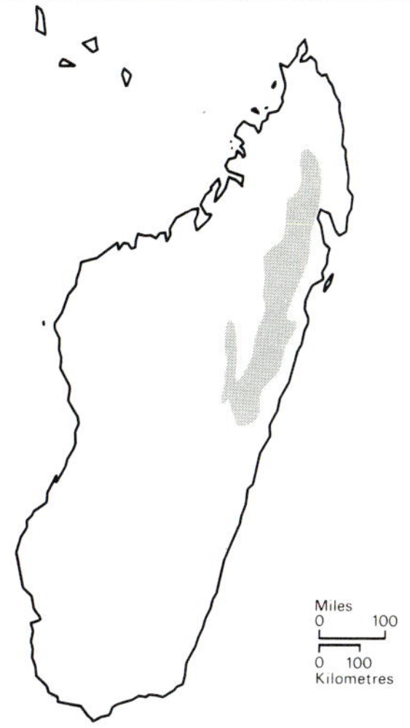

The indris GENUS *Indri*

A chorus of powerful barks echoes through the forest-clad hills of eastern Madagascar, stopping almost as soon as it begins. Nearly immediately, this is followed by a loud, but somehow plaintive, howling – like that of dogs, only stronger and more musical. The howling song, which is also in chorus, slowly dies away after a minute or so and the more familiar background noises of a tropical rain forest take over again. Branches creak in the wind, insects buzz and whine, and birds call in the trees. Then more barks and howls come floating across a valley, in answer to the first.

These doglike noises are the territorial advertisements of the largest living prosimian, the indris. Adult pairs of indrises divide up the forest so that each family lives in its own little patch, from which it can obtain the necessities of life. Like the avahis and sifakas, indrises depend mainly on leaves for the bulk of their diet and eat fruit when they can.

The adults weigh slightly over 10 kg (22 lb); and in their size, their leafy diet, their monogamy and their vocally advertised territoriality, they bear a remarkable resemblance to the lesser apes of South East Asia. Indeed, indrises are often described as the 'apes of the prosimian world'. They have even come close to losing their tails, being left with only rudimentary stumps.

As with most folivorous (leaf-eating) primates, their life styles are none too energetic. Although they bound quickly through the trees or engage in territorial border disputes from time to time, they seldom become active until the sun has been up for an hour or two, and bedtime is often two, or even three, hours before dusk.

During the day, they forage steadily, with the female and perhaps

The indris is the nearest thing to an ape to have evolved among the lemurs. It is the biggest primate living on Madagascar and its tail is reduced to a mere stump.

one or two of her offspring taking the prime feeding sites high in the forest canopy. Males defer to their strong-minded mates and do most of their feeding in the less desirable lower levels. Should a male be tempted to forget his lowly station in life, the female is always there to remind him with a sharp attack.

No animal matter is consumed, but on most days, the family will descend to the base of an uprooted tree and take it in turns to feed on the bare earth. They have favourite spots for doing this, and it may be that the earth contains some important mineral that is not otherwise available in their diet.

Relations within the family are not exactly stimulating. Individuals just busy themselves with their foraging or digesting during most of their rather short daily activity period, but once in a while one indris will groom another and then perhaps the roles will be reversed. Unlike sifakas, two indrises never groom each other at the same time, although teeth and tongues are used in much the same manner as by the former species.

Bouts of playful wrestling sometimes occur in breathless silence, but these tend not to involve the adults. Little aggression occurs other than in arguments over choice feeding sites, and the sex life of an indris is almost non-existent. The ages of the offspring indicate that sexual activity must occur every two or three years, but love-making is far from being a regular pastime.

In spite of their love of the quiet life, indrises manage to capture the imagination, and perhaps the awe, of those human primates who are lucky enough to encounter them. With their upright carriage, their thick, silky, black-and-white fur, and their big, yellow eyes that stare over prominent, black muzzles, there is an inescapable majesty about these peaceful creatures.

The aye-aye GENUS *Daubentonia*

If there is a prize for the weirdest primate, the aye-aye must be the odds-on favourite to win it: although if extinction leads to disqualification, the judges may have to hurry.

Whereas most Malagasy primates have 36 teeth, the aye-aye has only 18 and lacks the typical, prosimian dental comb. However, its teeth grow continuously throughout its life, a phenomenon that caused it to be classified as a rodent for many years. It has claws, rather than nails, on all of its digits except the big toes, and its skeleton provides a list of aberrant features that set it apart from other primates. Nevertheless, there is enough about it that is clearly 'primate' to place it in our order, even if it has to be given a super-family all to itself.

No simple analogy with another species is possible for describing

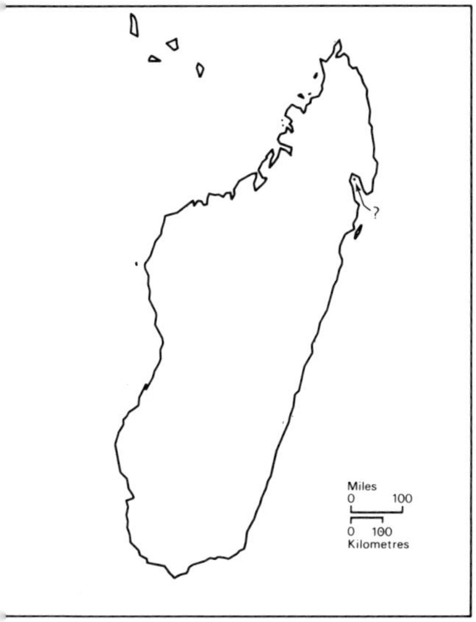

how the aye-aye appears in its natural state. Mobile Mickey Mouse ears set off a slightly catlike face, with big, staring eyes. The long fur is predominantly black, with white highlights and irregular patches showing through in contrast. The tail is longer than the body and more plumed than a fox's brush. From tip to tip, the average adult aye-aye is just under 1 m long (about 3 ft).

Add all this to its nocturnal way of life and it is not surprising that it has become an object of local superstition. Glimpsed occasionally by moonlight, it must be a hard creature to regard as entirely mortal. Unfortunately, some of the superstitions prescribe that it must be killed, although others are said to offer it the protection of a taboo against just that.

Aye-ayes are so rare and unstudied that the details of their private lives are little known. They spend the daylight hours in durable nests that they build high up in the trees. These shelters are usually constructed in tree forks where green twigs can be woven around convenient support branches. More twigs and leaves are added and the whole structure may last for years. Although aye-ayes are solitary beasts, it is thought to be possible that any one nest may be occupied by different individuals on different days.

When they are moving about in the trees, aye-ayes sometimes call to each other and they grunt when disturbed, but they are mostly rather silent – with one big exception: they are incredibly noisy eaters. A feeding aye-aye will slobber and chomp its way through a juicy fruit, munching up any insect larvae that it finds therein, even if it has to crunch up the stone within the fruit to get at them. Unripe coconuts are attacked with enthusiasm, being torn and gnawed into with the teeth until a hole is made via which the milk and flesh can be extracted. To reach at the latter, the aye-aye employs what is perhaps its oddest physical characteristic: an enormously long and thin middle finger that can be inserted deep into the coconut to scrape out the flesh.

Probably, the most important function of this finger is in grub-hunting. When doing this, an aye-aye walks slowly along (above or below) a branch, with its nose pressed to the bark. Every now and then it slows down, presumably when it has heard or smelled something interesting. It then cocks its head and bends its big ears towards the focus of its attention. The next stage is to gnaw furiously at the branch, discarding splinters of wood and pieces of bark until it has made an opening to the tunnel in which the grub is lurking. As likely as not, the quarry will be a wood-boring beetle larva. The aye-aye then inserts its lengthy digit into the tunnel and squashes the grub with it. The remains of the prey are then removed bit by bit and licked off the end of the finger. In many ways, aye-ayes are like woodpeckers, which listen for wood-boring larvae and then peck at them, and woodpeckers are absent from Madagascar.

Possibly the strangest of all primates, the aye-aye has long been the object of superstitious persecution by man and is now on the brink of extinction.

The aye-aye has already been thought to be extinct once, but it was rediscovered near the east coast of Madagascar in 1957. Even then, there was little doubt that the small population that had been found was on the way to oblivion unless something could be done to save it.

Arrangements were made to protect its habitat, but by 1966 it was obvious that more drastic measures were necessary and two French biologists spent weeks collecting nine protesting individuals which were duly released on the island of Nosy Mangabé. The idea was that they could be better shielded from disturbance there than on the mainland, although there is always the risk that wild animals relocated in this way will not take to their new environment for some unforeseen reason.

Nosy Mangabé is only about 5 km^2 (less than 2 square miles) in area, but it is covered in dense forest. For some years after the transfer, the only hope of the aye-ayes' survival was that they were hidden in the vegetation, for none was seen. Pessimism was the order of the day, but recent sightings of nests and of eyes that glow by torchlight suggest that all is not lost. The aye-aye may yet have a future.

II The Bushbabies and their allies: nocturnal leapers and creepers

The major prosimian family living today outside the confines of Madagascar is the family Lorisidae of Africa and Asia. All of its members are nocturnal, and they have adopted two rather different modes of life to cope with their nocturnal existence in the fiercely competitive continental world.

Members of the subfamily Lorisinae creep about in the trees and are represented by one lightly built and one more robust genus in each of the two continents. Lorisine primates never leap.

The subfamily Galaginae is found only in Africa, where the species have adopted a faster-moving way of life that involves specializations in leaping. There is only one galagine genus, but it has successfully distributed itself throughout most of the area south of the Sahara.

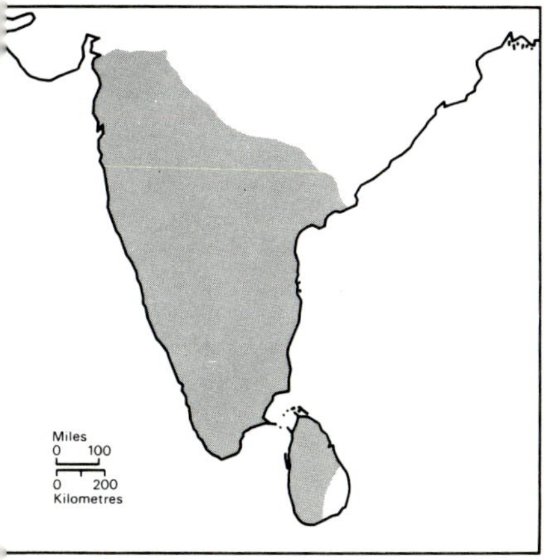

The slender loris GENUS *Loris*

No jerky movements betray the slender loris, which is both hunter and hunted, as it creeps about its business in the Asian night. With perfect fluidity of motion, each limb is brought forward in turn to grasp the thin branches of the forest understorey. Somewhat spindly in appearance, a slender loris is about 23 cm (9 in) along its head and body, and weighs about 300 g (10.5 oz). So small a creature

The slender loris creeps very deliberately through the forest on stilt-like limbs, but it can pounce quickly on edible insects.

must be very vulnerable to a host of nocturnal predators, including snakes, owls and climbing mammals, such as certain types of cats and civets. But most such predators detect their prey by its movement, so a slender loris always has a good chance of gliding by unnoticed, or of freezing until the danger has passed.

As a predator itself, it creeps up on a variety of small animal life, pouncing at the last moment to grab with both its hands and follow through with a sharp bite. Insects are the usual prey, often being snatched up as they take off, just a split second too late; but snails are also crunched up with relish, and eggs, sleeping birds and lizards are especially nutritious prizes.

In common with the slow loris, the potto and the angwantibo (all of which are members of the lorisine subfamily), the slender loris has a rather indiscriminate taste in invertebrates, snapping up those butterflies, caterpillars, beetles and millipedes that protect themselves from most predators by having noxious smells or tastes. It takes more than that to put off a hungry slender loris.

The slender loris is typical of nocturnal prosimians in conducting much of its communication in the language of odour. Females let off a strong vaginal discharge when they are ready for a male, although those in the vicinity will have probably been aware for days that a particular female was going to be receptive soon, because such messages are contained in the urine and left for others to find. It would be a profligate loris indeed that would relieve itself onto the forest floor, metres below. Instead, personal details of age, sex, condition and perhaps even individual identity are carefully deposited on the branches. Small quantities of urine are usually emitted at intervals as the animal moves forward, so that the maximum area may be marked with economy.

Slender lorises also 'urine-wash', a clever way of leaving a message far and wide by automatic printing. This consists of wetting themselves, drop by drop, on the palm of one hand, after which the palm is rubbed thoroughly on the sole of the foot on the same side of the body. The process is then repeated on the other side and henceforth until it dries, the urine message is left with every pace. This pattern of behaviour has another advantage in that it makes the soles of the hands and feet sticky, giving the lorises a better grip in the trees.

There is only a single species of slender loris, but different races appear to be adapted to different types of forest and woodland in India and in Sri Lanka. There is even a race that has developed extra-thick fur to combat the continuous damp and cool of the central Sri Lankan highlands. All races probably lead similar sorts of lives, creeping about in search of insects. They probably occupy much the same ecological niche as the angwantibo in Africa, but the details are still something of a mystery.

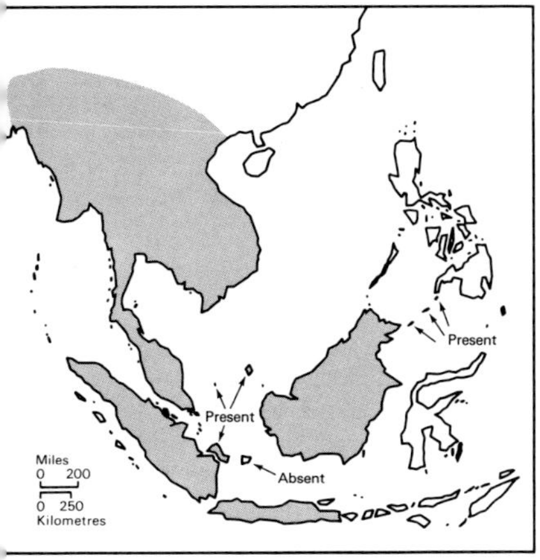

The slow loris GENUS *Nycticebus*

Throughout the jungles of South East Asia and the western islands of the Indonesian archipelago, wherever humans have not destroyed their habitat, slow lorises have their home. Yet widespread as they are, little is known about their way of life.

Like the African pottos, slow lorises are of the robust, heavier type of lorisine. They have to depend on fruit for the bulk of their diet because they need more than they can get by eating animal matter alone. Most races weigh about 1.4 kg (3 lb), but in Vietnam there is a pygmy race of about half that size. Possibly, these pygmies should be counted as a separate species, but nobody knows enough about them to be sure.

Slow lorises probably do not see all that well in the dark, but they live in a world that is characterized by many different fascinating smells and sounds. A slow-moving 'creeper' has less need of accurate long-distance vision than a rapid 'leaper' like a bushbaby.

Entirely nocturnal, slow lorises have a marked aversion to daylight and spend the hours between dawn and dusk rolled up into furry balls, sleeping hidden in dense foliage. When night falls, they set out in search of fruit, insects, eggs and other snacks, ignoring any bad odours with which prey species attempt to protect themselves, yet able to emit nauseating, thoroughly repellent smells if they themselves are attacked.

For additional protection, a slow loris has big black marks on its face which increase the apparent size of its eyes, so giving it a chance of scaring off a predator by staring at it. Large false eyes are a common protective tactic in the animal world, employed by creatures as diverse as butterflies, mammals and birds. No doubt, some predators learn to spot the deception, but it pays to be careful if you are thinking of biting another animal, and a moment's hesitation by a predator may mean a reprieve for its intended prey. It is not, therefore, surprising that the trick is used by so many species. If it does not work for the slow loris, it can always fall back on a breathy, whistling sound which, together with its steady, slow, almost gliding locomotion, is uncomfortably reminiscent of an aroused cobra.

Such deceptions are not born of intelligence; rather it is the case that vulnerable animals that happen to look fearsome in some way or other have survived better than their more innocent-looking fellows. Over time, the bluff has been improved and refined as those that are best at it leave more descendants in each succeeding generation.

From what little is known about the social life of the slow loris, it seems to be a solitary beast, at least when it is foraging. No doubt it keeps track of its neighbours by their smells and by occasional

OPPOSITE **This adult male slow loris is a creature of the night and best seen when the moon is full.**

meetings, but it could hardly be described as gregarious. It is given an early lesson in making do with its own company when, at one day old, it is parked by its mother when she goes off to feed. Baby slow lorises are born with their eyes open and able to cling to thin branches. When left hanging beneath a convenient support at night, they are utterly inconspicuous and the mother can be fairly sure of the infant's safety until she returns before daybreak.

Infants grow quickly and can walk – albeit rather unsteadily – at the age of two weeks. By the age of ten weeks, they become rather playful and in less than a year they can fend for themselves.

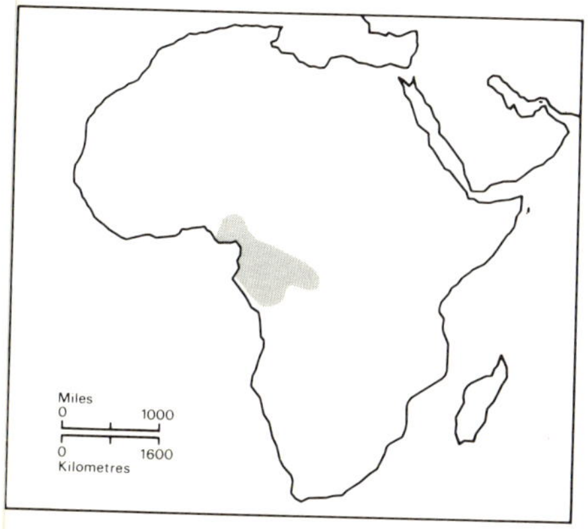

The angwantibo GENUS *Arctocebus*

Much of the land below the canopy of a tropical rain forest is relatively clear of undergrowth and no jungle knife or elephantine strength is necessary to walk between the trees. The millions of leaves high above are so efficient at blotting out the sunshine that no dense vegetation can survive beneath them. But, when lightning or the wind knocks over an ageing giant tree, then thousands upon thousands of little seedlings come to life and, revelling in the rare sunlight, they struggle upwards in a dense tangle of fresh, green growth. And, in so doing, they provide a home for the angwantibo, or golden potto, of central Africa.

Patches of forest that lack tree falls also lack angwantibos, for this spindly creature is a specialist at creeping slowly along thin lianas and small branches in the undergrowth, in search of its prey. Such places are rich in foliage, and here the nocturnal angwantibo finds the caterpillars and other insects that are the mainstay of its diet. So attached is the angwantibo to its milieu that it is incapable of climbing up big, mature branches and it will never cross more than a few metres of ground or branch that is open to the sky: a truly agoraphobic prosimian!

The tapetum, or reflective shield, at the back of the angwantibo's eyes shines in response to the camera's flash. Only prosimians have eyes like this.

Yet, for a somewhat etiolated, slow-moving mammal that weighs only about 210g (7.5oz), the undergrowth must have its dangers. The angwantibo comes right down to the ground quite often and is at risk from terrestrial as well as arboreal predators. For defence, it relies principally on not being noticed. Being something of a solitary creature and avoiding all conspicuous, sudden movements, its unobtrusiveness is probably enough to guard it most of the time; but when that fails, there is always an alternative stratagem to fall dack upon. This consists of arching its back and tucking its head between its arms so that its snout is placed under its armpit. In this position, it presents its attacker with a curved, pale brown or rusty-coloured furry body with a black patch of hair at its tail. If, as seems likely, the predator is attracted to this conspicuous rear end and bites it,

and divides most of the rest between fruits and gum, with an occasional bud or leaf for a change.

At a body weight of about 61 g (just over 2 oz), Demidoff's is by far the smallest of the bushbabies and ties with the grey mouse lemur for the title of smallest primate. In contrast, the thick-tailed bushbaby of eastern and southern Africa is the biggest galago species, sometimes weighing up to 2 kg (2.2 lb). The needle-nailed bushbaby weighs about 300 g (10.6 oz) and the other three species that are universally recognized all weigh about 260 g (9 oz).

Nails are good for backing up sensitive pads on the ends of fingers and claws are good for gripping smooth, broad tree trunks. Neither is very good at the other's task, but the needle-nailed bushbaby is an unusual species that manages to have the best of both worlds. A ridge runs down the centre of each fingernail and extends out beyond the finger into a sharp 'claw'.

When its fingers are relaxed, the 'claws' of a needle-nailed bushbaby lie flat behind its fingers and toes like ordinary nails. But, besides being a good leaper, it spends much of its time scurrying up and down big tree trunks and at such times the ends of the fingers are bent back so that the 'claws' dig into the tree. This opens up numerous arboreal routes that would otherwise be as impossible to use as a vertical sheet of glass.

In fact, the needle-nailed bushbabies are the most agile of all the lorisid species since, in addition to their trunk-scurrying practices, they can make the longest leaps of all. The leaping of bushbabies is often exaggerated, but the needle-nailed variety can certainly jump across a 5.5 m (18 ft) gap, as long as its trajectory can include a drop of 3 m or so. Such a leap is quite impressive for such a small creature. It may be that it gets some assistance from four small folds of skin that lie between its body and its limbs, since these are spread-eagled when it hurtles through the air and may provide a slight 'hang-glider' effect. When it jumps upwards between supports, however, the last thing that it needs is extra wind resistance, so it is appropriate that it does so with arms and legs stretched out fore and aft to produce a more streamlined shape.

The needle-nailed variety is really very closely related to the other bushbabies but some biologists would accord it the status of a separate genus. It is set apart by certain anatomical features, such as its teeth and the shape of its skull, but the living animal is, to all intents and purposes, very much a bushbaby.

OPPOSITE **Horsfield's tarsier of Borneo lives in a world of vertical supports. It clings to thin trunks with its hands and feet while propping up its body with its tail and scans the ground below for likely prey.**

3. THE ODD ONES OUT: THE DRY-NOSED PROSIMIANS

It is all very well making a neat division of the primates by categorizing them as prosimians or simians; and this works perfectly for 50 out of 51 genera. Unfortunately, there exists in South East Asia a group of small, nocturnal primates, called the tarsiers, that are quite obviously prosimians – until they are examined closely. They then turn out to have dry, simian-type noses, and their unborn babies are attached to the walls of their mothers' wombs in a way that is typical of the simian primates, but never found among the other prosimians. It is hard to see why the tarsiers should have these characteristics unless they have more recent common ancestry with the simians than is the case for the other prosimians.

Equally telling, the anatomy of a tarsier's eye clearly indicates that it has had a diurnal ancestor somewhere along the line; it has a focusing point in the retina (the light-sensitive part), known as a fovea. Now a fovea is something that is only useful in daylight, to which the rest of the retina has to be adapted. But the tarsier's retina is adapted for night vision, which clearly puts its fovea into the category of a useless inheritance. From the tarsier's point of view, it would have been better off if, like the other prosimians, it had inherited a light-reflecting tapetum behind the eye to enhance seeing in the dark. But it did not.

When we consider that all the other prosimians, even the diurnal lemurs, have nocturnal animals' wet noses for accurate smelling and fovea-less eyes with tapeta for night vision, it becomes fairly obvious that the tarsiers' ancestors must have gone through quite a long period of diurnal life to have lost these characteristics. Add to this their monkey-like wombs, and their closest relatives seem much more likely to be monkeys than lemurs, or lorises, or bushbabies. In consequence, the entire order of primates is usually divided into two suborders, the strepsirhines and the haplorhines. The former have a wet patch of skin around the nostrils, joined to the upper lip, whereas the latter have lost this in favour of hairy skin between the nostrils and the upper lip.

Thus the most fundamental division within the order does not correspond to that between the prosimians and the simians. Tarsiers bridge the gap by being prosimians with important simian characteristics – although they are certainly not simians, since the term refers to being monkeylike in form. Because of this, they have been given a section to themselves in this book. They are haplorhines that have gone back to a prosimian way of life. They and their close relatives were once widespread in Europe, Asia and North America and the tarsiers that are alive today are the last of a once great group.

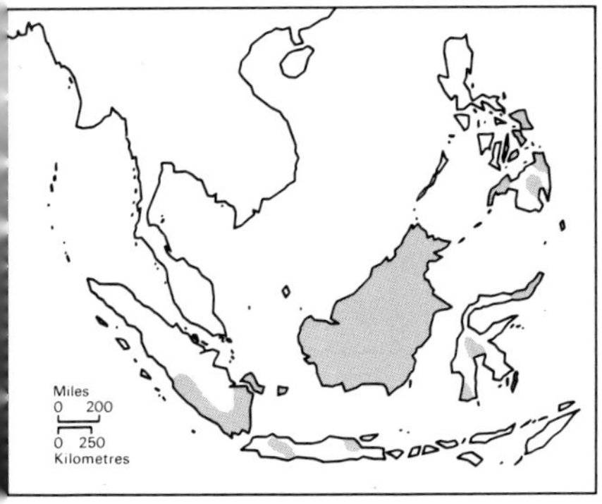

The tarsiers GENUS *Tarsius*

No matter what it stares at with its great, soup-plate eyes, a tarsier seems to regard the world with a mixture of astonishment and outrage. There are three species of these tiny and endearing creatures, with their short bodies, round heads, long and spatulate fingers and toes for clinging on, and powerful legs for kicking off. They are adept at clinging to thin tree trunks and can even sleep in this position, propped up by their long, nearly naked tails that they press against the trees.

When they move, they prove themselves to be superb leapers, able to cross gaps of 6 m (about 20 ft) with enough of a vertical drop. Yet they weigh only about 130 g (4.5 oz), and are therefore among the smallest of primates.

Little and charming they may be, but they survive by predation that is sometimes brave and fierce. Like owls, they move silently about in the night and drop quickly onto their victims. All sorts of large insects are eaten, and roosting birds are especially at risk. Usually, some sound will give away the presence of a potential prey and the tarsier will then fix upon it with its huge eyes, rotating its head through 180° if necessary, exactly like an owl. Its final attack is by leaping; and death comes with a few sharp bites. Even highly poisonous snakes are taken with cool efficiency.

For illustration of Horsfield's tarsier (see page 65).

Unlike many primates, a tarsier is something of a meticulous eater, consuming every bit of edible material, even down to the feathers of birds, and sometimes their feet and beaks as well. This is probably because a predator cannot afford to be profligate with its food and the wasteful, messy eaters of the primate order usually subsist on a vegetarian diet. It is much harder to catch a beetle or a bird than it is to pluck a flower or a fig, and the tarsier eats only the live prey that it catches.

The social life of these animals remains something of an enigma, but it seems that, on the island of Borneo, they probably live in monogamous families, whereas on the Philippine Islands bigger groups may be the norm.

As would be expected for a creature that is active by night, sounds and smell play the major roles in communication. Most of the tarsier's calls are extremely high-pitched, many of them containing elements that are beyond the reach of human hearing. On the other hand, their scent messages are quite strong and well within the range of the human nose. Some are made by marking certain places with urine, as do many prosimians, but their meaning is still a mystery.

OVERLEAF **The golden race of lion tamarin is among the most beautiful of primates. If the destruction of their native forests continues, all the lion tamarins may soon be extinct in the wild. Brazilian and American zoologists are trying to save them by breeding them in captivity (see page 81).**

4. THE MONKEYS OF THE NEW WORLD

The continent of South America is nearly 18 million km^2 (7 million square miles) in area and extends from approximately 12°N to 54°S. It is broadest just south of the Equator, and about three-quarters of the land is within the tropics. The most important physical features are the high Andean mountain chain that runs north-to-south along the western edge of the continent and the massive lowland areas with their great, flat river basins that lie to the east of the mountains. There are also some outcrops of high land in the east, particularly in Brazil and Venezuela, but very few of these exceed 2000m (6560ft) above sea level. In contrast, large areas of the Andes soar above 4000 metres (13,120ft).

The great Amazon river basin stretches from the foothills of the Andes in Colombia, Ecuador, Peru and Bolivia across Brazil to the Atlantic Ocean. Covering about 7 million km^2 (more than 2.5 million square miles), its area is larger than that of the whole of Europe and contains the biggest tropical rain forest in the world.

South America is connected to the larger North American continent by the narrow Isthmus of Panama and in order to understand its primates it is important to realize that they evolved from a common ancestor in South America itself, although today some species are to be found in the rain forests of Central America and Mexico. It is generally believed that most of South America was covered with forest until relatively recently, although it is not possible to be specific about the ages of the grasslands that at present exist in southern, central and eastern Brazil. Currently, human influence is favouring the rapid spread of drier, open country at the expense of the humid forests. What is clear is that South American animals are predominantly adapted to live in the forest and there has been no proliferation of grassland and woodland forms as has happened in Africa with, for example, the formation of vast herds of ungulates.

This rule holds true for the primates, all of which are adapted to an arboreal way of life. South America has no equivalent of Africa's semi-terrestrial baboons or apes. It has an array of 16 genera of monkeys, no apes and no prosimians.

The origin of these New World monkeys is obscure, largely because of the incomplete nature of the fossil record and because of differences of opinion about the relative positions of North America, South America and Africa millions of years ago. The earliest record of monkeylike primates – in fact, of any primates – in South America comes from about 40 million years ago, when the continent was certainly surrounded on all sides by ocean.

One school of thought holds that these animals came from North America, which is known to have had resident prosimians as long as 60 million years ago. They are presumed to have reached South America by rafting along a chain of volcanic islands that provided a plausible link between the two great land masses.

The alternative school of thought rejects this theory on the grounds that the North American prosimian fauna did not contain anything that was sufficiently monkeylike to be the ancestor of the New World monkeys and the many similarities between the simians of the New and Old Worlds point to a common, monkeylike ancestor. This school theorizes that the first South American primates rafted across a few hundred miles of ocean from West Africa, pointing out that the Atlantic was a lot narrower 40 million years ago than it is today, and arguing that a much greater extent of open ocean separated the two American land masses.

North and South America became joined about 5 or 6 million years ago, since when there has been a considerable exchange of animal species. However, the monkeys' invasion (or re-invasion) of North America has got no further than the northern limit of the Mexican rain forest; and the ancient prosimians of the north have long since died out.

The New World monkeys of today are all included in the superfamily Ceboidea but there is much argument about their classification at the family and specific levels. Their division into the three families Callitrichidae, Callimiconidae and Cebidae may well prove to be incorrect as new studies provide more information, but it has the merit of being simple, comprehensible and widely used, whereas the alternative schemes are, at best, not proven. Suffice it to say that primatologists have a gnawing sense of unease about ceboid taxonomy but have not yet agreed upon what to do about it.

In comparison with the Old World monkey species, the natural history of those of the New World is not well known. Not surprisingly, some of the biggest gaps in science's knowledge of them concern the ecology of those species that are rather small and difficult to follow in the forest.

I The Marmosets and Tamarins: little squirrellike monkeys with claws

The smallest of the South American monkeys belong to the families Callitrichidae and Callimiconidae. Apart from their size, the most obvious physical feature they have in common is claws on all digits except their big toes, which have broad, flat nails.

All species are diurnal; and their size and manner of running quadrupedally in the trees are reminiscent of squirrels. Twin births are common and the fathers are much involved in caring for the babies. Adult males and females are of the same general appearance and size, and it may be difficult to distinguish between them.

The term 'marmoset' is sometimes applied to any species of this group but it should really be restricted to those of the genera *Cebuella* and *Callithrix*. The genera *Saguinus* and *Leontopithecus* are the tamarins. The fifth genus, *Callimico*, is something of an odd-one-out and may not be at all closely related to the other four. For this reason, *Callimico* is given its own family, Callimiconidae, and the others are all assigned to the Callitrichidae.

Marmosets may be distinguished from the rather similar tamarins by the length of their lower canine teeth. Marmosets are 'short-tusked', having canines that are about the same length as their lower incisors, whereas tamarins are 'long-tusked', with canines that project up above the rest of their lower teeth. The 'long-tusked' condition is more typical of monkeys and apes and there is a major ecological significance to the difference between the two conditions. The marmosets' even row of incisors and canines is adapted for gouging holes out of tree trunks, as described in the following section. Although tamarins may chew wood from time to time, it does not appear to be a very important activity for them.

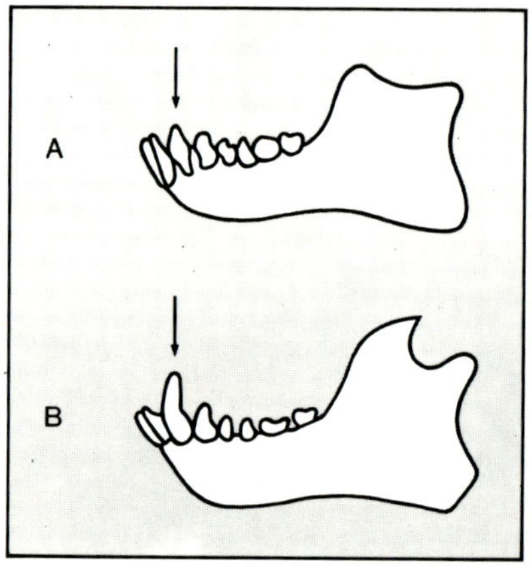

Lower jaws of a marmoset (A) and a tamarin (B). Note that the canine teeth (arrowed) project out much further in the case of the tamarin than in that of the marmoset.

The pygmy marmoset GENUS *Cebuella*

The smallest monkey in the world weighs about 125 g (4.4 oz) and may be found in the upper part of the Amazon basin, where it typically lives in families of about five or six individuals. It would, however, be remarkably easy to miss since, as befits one that is so small and vulnerable, it is an expert at concealing itself. Its grizzled brown fur, flecked with lighter and darker shades, is almost impossible to spot in the dappled forest light, and its movements come in sudden bursts between periods of frozen immobility. Like a lizard, its quick dash along a branch may be spotted at the corner of the predator's eye, but in the instant that it takes to turn and focus, the movement has stopped and the pygmy marmoset has effectively disappeared, even if the predator is staring straight at it.

At times when a quick dash might be perilous, the pygmy marmoset 'oozes' slowly out of sight, going either forwards or sideways with no sudden movement to catch the eye, until an intervening branch or trunk can offer some protection. Since it is too small to be of interest to human hunters, it is probably at greatest risk from birds of prey, so it tends to avoid the upper canopy layers of the forest where it would be most exposed; and at night it moves some distance from its feeding trees and finds security in a tree hole.

In spite of the hazards of its existence, this is quite a successful species and it has even managed to come to terms with human invasions of its forest. It probably gains some benefit when destruc-

South America's pygmy marmoset is the smallest monkey in the world.

tive modern man wipes out the forest birds of prey by introducing agriculture, and has shown itself to be able to adapt to changes in its habitat. Its most important habitats are riverine and flood-plain forests, but it manages to survive in the small areas of rather scrubby forest that are often left on farms and even ventures a little way into pastures to catch grasshoppers. Flexibility seems to be its watchword, and various new species of tree that are introduced by man are experimented with, and utilized for food if suitable.

The pygmy marmoset is fairly omnivorous in its tastes, taking many types of insects, spiders, small fruits and buds and an occasional lizard. However, there is one item it does not seem to be able to do without, and that is the exudate it obtains from a variety of trees by making and maintaining a large number of holes in the bark. Pygmy marmosets live in monogamous families, each of which inhabits a home range of less than one third of a hectare (three-quarters of an acre), which nevertheless contains several large trees that are covered with these small, disc-shaped holes. The marmosets gnaw into the bark with their lower front teeth, possibly taking days or weeks of repeated visits to develop any particular hole to the depth at which something nutritious is exuded. Nobody knows whether it is gum or sap that comes out, which is why the neutral term 'exudate' is employed, but whatever it is, it is so important that pygmy marmosets spend about 70 per cent of their feeding time on it. Furthermore, their local distribution in undisturbed rain forest seems to be dependent upon the presence of trees that are big enough and of the right species to be tapped.

Exudate holes are generally concentrated low down on a tree trunk but may come to cover an entire trunk over a long period of time. The marmosets have to be skilled enough to excavate to just the right depth, and no further, so that the hole will yield a slow, resinous ooze that can be harvested at regular intervals. Over-enthusiastic digging might kill the tree. On a typical day, a pygmy marmoset spends the first two hours or so of daylight collecting exudate and feeding on other items, and only in the middle and later parts of the day, when feeding is not so urgent, does it get on with the longer-term investment of scraping out more holes.

The rewards of exudate-tapping lie in the high sugar and mineral content of both sap and gum, making them reliable sources of energy and trace elements in the diet. It seems likely that pygmy marmosets gain sap from some trees and gum from others, depending on the species that is involved. The important point is that the supply is fairly constant, whereas fruit and insects may be unobtainable in some seasons.

There is only a single species of pygmy marmoset, but it is very closely related to the true marmosets and these also are exudate-tappers. However, pygmy and true marmosets are never found together in the same geographical area.

The true marmosets GENUS *Callithrix*

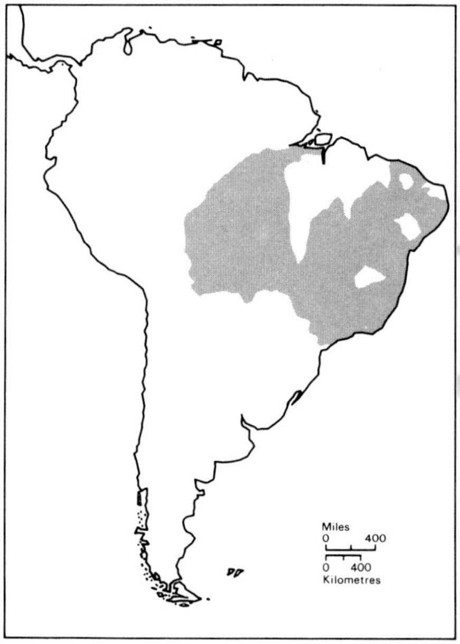

Like their diminutive cousins of the preceding genus, true marmosets organize their lives around their families. Among common marmosets, neither the male nor the female of the pair will tolerate rivals for the attentions of their mate, and between them they are extraordinarily efficient at rearing large numbers of young. Where larger primates may rear one infant every year or two, or at even longer intervals, marmosets usually produce twins and are as likely to have triplets as singletons. Not only that, it is routine for the female to give birth twice a year.

This production line begins with a gestation period of about 140 days, at the end of which the female gives birth, usually at night, only two or three hours after the first contraction. The newborn marmoset weighs only 30g (1 oz) but it must be able to cling immediately, for its mother gives it no help as it struggles out of the birth canal, waving its arms around until it makes contact with her fur. If it falls and hangs suspended by the umbilical cord, she will only reach for it if it has the energy to make a tiny cry of distress. Healthy infants usually haul themselves onto her fur and struggle with remarkable accuracy to a nipple within less than 20 minutes. Clearly, it is a disadvantage to be born third of three since the female has a single pair of nipples and the third animal will not get a look in for the first few hours. The first sign of real maternal interest usually comes after an infant has reached a nipple, at which point the female licks its face and it opens its eyes for a look at the world.

The placenta is usually eaten by the mother, and the father and older siblings show little interest in the birth itself. However, they do soon take a protective interest in the new arrivals and within a few hours they take over some of the carrying duties. The mother and father have most work to do for the first two sets of offspring, but after that the firstborn are usually present to help out, being nearly adult themselves.

Marmosets grow up quickly: they take their first interest in solid food only four or five weeks after they are born, sometimes chewing on insects that are held in the mother's mouth, or eating items that have been premasticated by the father. The mother makes sure that weaning occurs shortly thereafter, although other members of the family are more indulgent and will carry infants until the seventh week; by the twelfth week, the infants are almost independent but they return to sleep on their parents' backs until they are about four months old. At the age of 14 months, they reach full physical maturity, with an adult weight of approximately 370 g (13 oz).

In their family life, development and adult size, common marmosets are probably typical of the genus, although not much is known about many of the races or species. It is not even certain how many species of true marmoset there are, the real number lying between three and seven. The problem is that although some marmosets are well known in captivity, they are difficult to study in the wild, and it will take years of effort to sort out the behavioural, ecological and physiological differences between the 12 or 13 different-looking members of the genus.

With any type of animal, it can be difficult to distinguish between, on the one hand, greatly differing races of a single species and, on the other hand, very closely related species. As explained earlier, members of separate species do not leave common descendants (unless man interferes) and the problem with the marmosets is that they are all rather similar in many ways and nobody knows how much interbreeding there is between certain varieties. The situation is fast becoming impossible to sort out as habitat destruction creates artificial barriers between some populations and brings others together where they would not normally have contact.

There is still much to be learned about the everyday life of marmosets in the wild but their diets are broadly similar to that of the pygmy marmosets, consisting of insects, spiders and fruit, with a considerable dependence on tree exudates. They are found in a variety of forest types from undisturbed primary forests, through riverine forest and scrub in more open country, to regenerating,

The white face and ringed tail are typical of the white-fronted marmoset.

young forests in areas that have formerly been cleared, and even quite artificial habitats such as orchards and gardens. They appear to do best in young growth where there is dense vegetation with many creepers and shrubs. But, wherever they are found, marmosets give away their presence by peppering certain trees with holes. They not only harvest the exudates from these holes; they also scent-mark them by means of urinating into them and rubbing their genitals onto the wood nearby. Since it is the dominant animals in any group (usually the adult pair) that do this most often, it is almost certainly a way of establishing ownership of the area. How-

ABOVE LEFT **Like all the marmosets and tamarins, the buffy-tufted-ear marmoset has claws rather than nails on its fingers and toes, with the exception of its big toes which have broad nails.**

ABOVE RIGHT **There are three races of silvery marmosets that range in colour from dark all over to pure white with a creamy tail. In this variety, the white body contrasts sharply with the dark tail.**

ever range use and territoriality have yet to receive detailed study.

While it is difficult to follow small, quiet monkeys as they go about their daily business in the forest, it is perhaps surprising that common marmosets are not better known in view of the fact that they are one of the commonest primates in use in biomedical laboratories. However, under such circumstances, their significance usually lies in the ease with which they breed in captivity and in similarities between their physiology and that of humans, especially where their reactions to drugs or diseases are concerned, and in these conditions their natural history is of little consequence.

The tamarins GENUS *Saguinus*

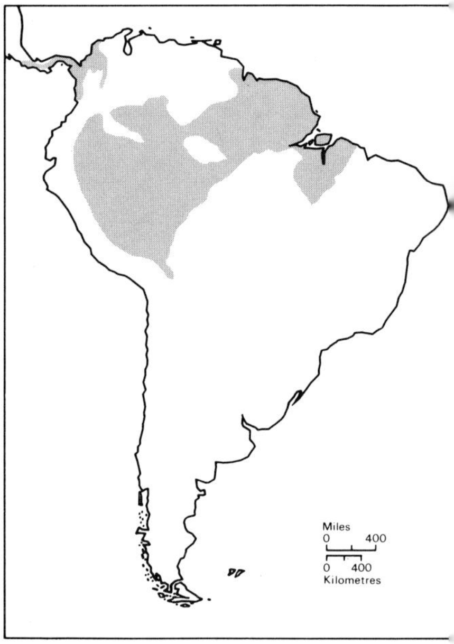

Soft, high-pitched chirps and twitters and an occasional flash of a squirrel-like animal running along a branch are all that would usually betray the presence of a group of white-footed tamarins in northern Colombia, or saddleback tamarins in Peru, or midas tamarins in eastern Brazil, or any of the 11 species and 30 or so races of this successful genus. Tamarins are distributed from the Amazon basin to the Isthmus of Panama. They are the only callitrichids to have spread north out of the great, moist Amazon forests, across the highland barriers of central Colombia and into the drier forests beyond.

In Panama, the only Central American species, the rufous-naped tamarin, is probably fairly typical of tamarins in general in most aspects of its life. Although the various species and races show great variety in the patterns and colours of their fur, and in the extent to which they have moustaches or smooth faces or flowing manes, they are all of similar size and their formula for survival seems to be similar wherever they live.

The rufous-naped tamarin has a rather bald-headed appearance with crinkly ears rather like a guinea pig's. Adults weigh about 450g (1 lb) and have a basically monogamous life style, albeit one that seems to permit liaisons outside the 'marriage' and perhaps the occasional 'divorce'. To a biologist, this is slightly surprising, because when mammals practise monogamy, they usually do so in a rather more wholehearted fashion. Among the primates, gibbons provide the best example of this. However, rufous-naped tamarins live in groups of about three to nine animals in which the core of the group is based upon a single adult pair and their offspring. In addition, there are more transient individuals present: these are usually adolescents or young adults that have joined the group from elsewhere. They seem to form a mobile, non-breeding section of the population, perhaps living temporarily with successive groups and biding their time until the opportunity arises to become one of a mated pair.

The surprising thing about tamarin groups is that only the one pair have offspring, even though several sexually mature individuals may be present and the male of the pair is likely to copulate with females other than his mate. It would be interesting to know what sort of contraception is involved. As among marmosets, the father and older siblings play important roles in caring for the babies, from the first few hours of their lives until they become independent.

This type of social organization is possibly typical of many, if not all, tamarin species and may well occur among the marmosets. The 'divorce' element certainly occurs among Colombia's cottontop tamarins, where one or other of the mated pair sometimes strikes out on its own to try a new group.

Tamarins may be found in a variety of forest types, from tall, undisturbed equatorial rain forest to smaller-statured dry forests, areas of scrubby regrowth, and even cultivated land where they can raid orchards on forays out of the fallow patches that they use for refuge. They are most successful where dense tangles of young trees and creepers give them an advantage over the larger monkeys which have more difficulty in moving among the small, flexible branches. Dense vegetation also offers the tamarins protection from birds of prey.

Nevertheless, the moustached tamarin provides an example of a species that forages in the upper layers of undisturbed rain forest, as well as exploiting the more typical tamarin habitat. Perhaps it is because of this that moustached tamarins are able to share some patches of forest with saddleback tamarins. Groups of the two species often join together for a few hours, or for a day or so, travelling and feeding quite amicably. On such occasions, the moustached tamarins tend to lead the way and to feed rather higher up than the saddlebacks, so the former probably do not suffer too much loss of food by the association, and perhaps the latter benefit from having extra lookouts up above.

Fruits, insects and spiders form the bulk of a tamarin's diet, although they also eat buds, soft young leaves and an occasional vertebrate, such as a small frog. The amount of land that each group occupies depends very much on the type of forest in which it lives and the forest's richness in terms of food production. In addition, a group can live in a smaller range if it can exclude competing groups of its own species, than if it is unable to do so. The rufous-naped tamarin provides a good example of this, with groups defending exclusive territories of about 26 hectares (64 acres) in lowland forest, but occupying overlapping ranges greater than 32 hectares (80 acres) – 23 per cent larger – in forests at a slightly higher altitude.

Like the common marmosets, many tamarin species are far better known in zoos and biomedical laboratories than they are in the wild. Colombia's rare cottontop tamarin has been one of those favoured by laboratory scientists. Sadly, the continuous expansion of human settlement in northern Colombia has had the twin effects of destroying the cottontop's habitat and bringing it into contact with animal collectors who can readily sell it in the port cities, either for export or for the local pet trade. Its small size and attractive white plume of hair on its head make it a favourite object of human affection, even though for every pet that is sold in the market many die during the trauma of capture and transportation. As long as people are prepared to buy pet monkeys, there is probably much to

ABOVE LEFT The cottontop tamarin is a rare species found only in Colombia. Because it makes such an attractive pet and can also be used in laboratories, its numbers in the wild have been much reduced by indiscriminate collecting.

ABOVE RIGHT This is the black-handed race of the midas tamarin. It comes from Brazil, near the mouth of the Amazon.

RIGHT The golden-headed race of lion tamarin is seldom found outside Brazil.

be said (from the monkey's point of view) for having a mean, ugly or thoroughly smelly disposition. Luckily, what was the biggest market for pet monkeys, the United States, has now prohibited their importation; and since the latter part of the 1970s, most South American countries have banned primate exports altogether.

The lion tamarins GENUS *Leontopithecus*

Early in the sixteenth century, when the first Portuguese settlers arrived in the Atlantic coast region of what is now southeastern Brazil, they found vast tracks of rain forest stretching far inland. Today, only a few isolated pockets of that forest remain and the area has become the major industrial and population centre of modern Brazil. Rio de Janeiro, São Paulo and other cities spread inexorably outwards; and away from them, farmland dominates the countryside. Small wonder that the local primate populations are severely endangered.

The single species of lion tamarin has probably never had a very wide distribution and today, it is touch-and-go as to whether it will survive at all in the wild. There are three distinct races of lion tamarin, all of which have the characteristic, leonine manes surrounding their rather bare faces. Each of the races is found in a separate part of the region, so that they have no natural contact with each other and the species is in danger of becoming extinct in the wild, one race at a time.

The golden lion tamarin is a startlingly beautiful creature with brilliant golden-yellow hair. In common with the other two races, adults weigh about 600g (1.3lb) and favour the middle layers of undisturbed, mature forest for their habitat. They live in family groups and are dependent for shelter upon the hollows that are found in the trunks of old trees. They retreat into these hollows at night, using narrow doorways that are too small to admit any predator that would be large enough to kill them. Favourite tree holes may be used for years, and they actually become more comfortable inside as cushions of golden hair gradually build up at the bottom.

Younger forests, of the type that grow up after the original vegetation is felled, offer fewer such shelters and therefore make poor homes for vulnerable little monkeys, even if they can offer sufficient insects, spiders, fruit and birds' eggs to feed them. Nevertheless, the destruction of the old forests has forced many of the remaining lion tamarins to try to survive in the areas of younger growth. It is perhaps an ominous sign that the golden lion tamarins in today's secondary forests are smaller than those that were found in yesterday's primary forests, and their hair is less shiny.

For illustration of lion tamarins (see pages 68–69).

The golden-rumped lion tamarin was once thought to be extinct. It is still the most endangered member of the species.

The two other races are predominantly a rich black colour with contrasting patches of golden hair, mainly where suggested by their names; they are the golden-headed and the golden-rumped lion tamarins.

Goeldi's monkey GENUS *Callimico*

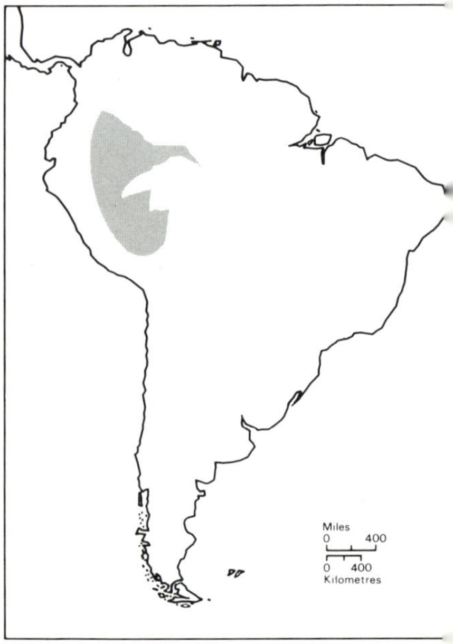

A little, pug-nosed, black monkey that comes from the upper Amazon basin has given primate taxonomists something they can really argue about. Superficially, it is a typical member of the callitrichid family, being rather squirrel-like and about the same size as a tamarin. It weighs about 500 g (1.1 lb) and its long, silky pelage, with something of a mane around the head, seems just right for the family. It also has the normal complement of claws on all its fingers and toes, except the big toe, which has a nail, as among the marmosets and tamarins.

However, it also has several characteristics that set it apart. A close examination of its anatomy reveals some features that are clearly like those of the callitrichids, some that are more like those of the other South American family, the cebids, and some that are more reminiscent of tarsiers. Anatomists use such things as fine details of bone and muscle structure, or the exact form of the tongue or genitals, as evidence of relationships. This is because they need to look for patterns within families or genera where they can distinguish between those characteristics that have been recently modified by natural selection upon the immediate forebears of the animal in question and those that have been inherited virtually unchanged from some ancient ancestor.

Although superficially like a marmoset, Goeldi's monkey is in many ways intermediate between the two major families of American monkeys, the Callitrichidae and the Cebidea. For this reason, it is put in its own family, the Callimiconidae.

In the case of Goeldi's monkey, the cheek teeth are considered to be of great diagnostic importance. The callitrichids all have five cheek teeth, but Goeldi's monkey has six. This is a compelling piece of evidence for different ancestry a long way back, since it is otherwise difficult to see why monkeys that lead such similar lives should have dental arrangements that differ so fundamentally. In addition, it gives Goeldi's monkey something in common with the larger South American genera, all of which have six cheek teeth.

The reproductive biology of Goeldi's monkey also sets it apart from the marmosets and tamarins, since it typically gives birth to one young at a time. The father shows great interest in his offspring from the beginning of its life, but it is not until the third week, when the mother is beginning to show signs of needing a baby-sitter once in a while, that he actually carries it around. From then on, he carries it on his back most of the time, returning it to the mother when it is hungry. Infants rapidly learn to be cooperative when being transferred from one parent to the other.

When danger threatens, parent Goeldi's monkeys hastily hide their infants in low vegetation before running away, a pattern of behaviour that is extraordinary for a monkey and even a little reminiscent of those prosimians that park their babies out of harm's way while they are feeding. Although monkeys and apes do not generally go in for the practice, many other animals hide their babies in some way or other for their safety; but it is impossible to say whether Goeldi's monkey has inherited an ancient emergency procedure from a primeval ancestor or re-invented an old idea.

Fleeing adults take care of themselves by zigzagging rapidly as they run and leap away. This probably helps to confuse predators, and the fleeing monkeys keep to the lower layers of the forest where they are hidden from birds of prey but high enough to be out of reach of such dangerous beasts as ocelots and other small cats.

Groups of Goeldi's monkeys are probably based on the family unit, perhaps with some other individuals who join in, much as is the case among the tamarins. Usually, there are six or seven monkeys in a group. Like many of the callitrichid species, they are most often found in and around bamboo patches in rather scrubby areas where the vegetation is relatively young and dense, but they also seem to occur in older, undisturbed forests. In the latter case, they still probably gravitate to those places where there is most undergrowth. Whatever the type of forest, they spend most of their active time quite close to the ground. When night falls, they climb up the trees to a height of about 10 or 15 m (30 to 50 ft) and find some sheltered spot in which to bed down.

Like the tamarins, Goeldi's monkeys probably depend on fruit, insects and spiders for most of their food, and they are not averse to such small vertebrates as lizards and frogs.

II Latin America's Larger Monkeys: complex societies and prehensile tails

The remaining 11 genera of South and Central American monkeys are lumped into the single family Cebidae, which forms something of a mixed bag. In the progression up the primate ladder, these are the first of the really typically monkeylike forms to be encountered. They are long-limbed, agile at climbing by means of holding branches, and generally skilled at handling objects. Like the Old World monkeys, apes and humans, they have flat nails rather than claws on the ends of all of their fingers and toes.

Four of the eleven genera have 'palmlike' pads of bare skin at the ends of very mobile prehensile tails and a fifth genus can also hang by its tail in the trees. No other monkeys have such truly prehensile tails as these, but it is not a characteristic of all the family. One genus even consists of monkeys with short tails.

The very mixed nature of the family is reflected in the fact that its 11 genera are allocated to no less than 7 subfamilies. Their social organizations are, however, as diverse as might be expected from such a motley collection, with monogamous families and various types of larger groups, including some that divide up and rejoin in complex ways.

The squirrel monkeys GENUS *Saimiri*

In spite of their name, the squirrel monkeys are much less squirrel-like than are the marmosets and tamarins. Although very agile and highly arboreal, they run along the branches with more obvious limb movements and do not scurry and leap in quite the quick, nervous fashion of the callitrichid monkeys. They are also somewhat bigger: adult females weigh between 500 and 750 g (1.1 to 1.6 lb) and males between 700 and 1100 g (1.5 to 2.4 lb).

Although they are widespread in South America in almost every conceivable type of forest, and well-known in Europe, Japan and North America as pets and laboratory animals, there is still no agreement as to the correct number of species. All varieties of squirrel monkey are superficially quite similar, varying mainly in the shade of the coat and the shape of the white facial mask, but it is quite possible that different populations have very different physiologies. For simplicity, they may all be thought of as belonging to a single species, although many zoologists would want to separate a red-backed, Central American form from those that occur in South America itself.

In the large forests of the Amazon, squirrel monkeys sometimes aggregate into what are some of the biggest groups of non-human primates to be found anywhere. There are reports of 500 or more monkeys travelling together, although these may not have been permanent social units. Most groups probably comprise fewer than 100 individuals; and where they are found in smaller forest patches, 20 or 30 animals is much more usual.

There is no doubt that it is the adult females that form the core of the group. They clearly relish each other's company far more than that of the males, and for several months of every year they are busy caring for the latest crop of infants. Those females that do not have young of their own to look after generally form special relationships with certain mother–infant pairs and spend time playing the role of an enthusiastic 'aunt'. For the first two weeks of its life, a baby squirrel monkey remains rather inconspicuous as it does little more than cling to its mother and alternate between sleeping and feeding. During this period, the other group members show little interest in it; and even the mother seems to ignore it for much of the time.

However, after this, the infant starts to attract the attention of all and sundry by moving about on its mother's body, and it is at this point that it often acquires an 'aunt', or even several of them. Most of these relationships are quite casual and temporary: they are based on a general female interest in the little ones which is manifested in approaches and attempts either to lift the baby off the mother or to persuade it to climb onto an appropriately positioned shoulder or back. Juvenile females that do this often frighten the baby by being too playful, and its indignant 'peeps' result in rapid retrieval by the mother. Older females show more maternal competence and it is at this point that a childless one will very often initiate a relationship that usually lasts for about two months. In doing so, she relieves the mother of some of the burdens of parenthood and provides more security for the infant.

The mothers tend to show most concern over their infants before a particular aunt is established, at the stage when other females begin to 'borrow' them; and they have more reason to worry when, shortly after this, the infants show signs of wanting to make their first independent explorations of the environment. Clearly, this is a hazardous period as it is so easy for an infant to fall out of a tree. Mothers vary greatly in the amount of protectiveness they show, but by the time their offspring have reached five or six months old, maternal patience is usually showing occasional signs of wearing thin and weaning is under way.

The infants themselves show more interest in forming play groups with fellows of their own age, and in making general nuisances of themselves by chasing and shrieking at other group members. This trend continues and, in the second year of their lives,

LEFT **Squirrel monkeys are widespread in the New World and there are several different varieties. It may be that they will eventually be classified into a number of different species, but the necessary studies have yet to be made.**

OPPOSITE **Flowers are not a major part of the squirrel monkey's diet, but they are relished on occasion.**

the males become noticeably rougher than the females and introduce sexual elements into their play. For their part, the females show increasing disenchantment with their male peers and develop their interests in the company of adult females and their latest offspring. They become adult and ready to mate at about 2½ years. The males, on the other hand, enter a transitional 'subadult' phase, during which they continue to play among themselves and with the group's younger juveniles. They also show an increasing amount of sexual interest in the adult females, although they will be four or five years old before they look and behave like mature adults themselves. They have a good chance of being driven out of the group by the adult males before this happens, in which case they presumably have to join up with another group, but it is not known how this happens.

Throughout this growing-up process, the adult males' most typical paternal contribution is to threaten the more irritating of the juveniles. For much of the year, the adult and older, subadult makes keep some distance between themselves and the rest of the group, although they do not actually move off on a separate course.

Squirrel monkeys are seasonal breeders, giving birth during the wet season, and the mating period sees physical as well as behavioural changes among the males. They put on layers of fat on their upper torsos, arms and shoulders, gaining as much as 20 per cent more body weight in the process. In addition, their testicles produce viable sperm – which is not the case at other times of the year – and they become very excitable for the three or four months during which they are sexually active.

This excitement is often rather counter-productive because the majority of the females and juveniles are much given to interfering with the activities of courting couples. Success, therefore, comes mostly to those that are discreet about their activities, even if they have to spend quite a long time gradually getting themselves left behind by the rest of the group. Noisy and impatient males invite interference time and time again, however willing their consorts may be to put up with their lack of subtlety. Courtship is brief and the whole affair takes anything from 5 minutes to $2\frac{1}{2}$ hours, after which there is no special relationship between the two animals concerned. Within their groups, squirrel monkeys are promiscuous.

Although relations among the females are generally easygoing, those among the males are more obviously aggressive. On occasion, they fight fiercely and they maintain a hierarchy of dominance among themselves, so that each one knows whom he can boss about and who can boss him with impunity. This dominance hierarchy may be detected by noting incidents of 'penile display'.

This behaviour is highly ritualized. A dominant monkey approaches a subordinate and makes various little growling noises. He then faces the subordinate and may place his hand on his head or his back, after which he bends his knee outwards and pokes his erect penis in the general direction of the subordinate's face, sometimes urinating slightly as he does so. To make his point more thoroughly, the dominant male sometimes bumps into the subordinate or even ends up by climbing onto his back.

For all their posturing, the males at the top of the penile display hierarchy do not necessarily have the most success with the females. Often they are precisely those who make so much noise and general disturbance when they are courting that the other females never leave them alone to get on with it.

That being so, it is not clear what biological advantage a male achieves from his social dominance. Perhaps they are more likely to be left alone when they have found a particularly good source of food. Insects are very important in the diet of squirrel monkeys and they often require a considerable amount of searching for, and it takes some skill to catch them on the wing.

Largely because of these factors, fruit forms the biggest proportion of a squirrel monkey's diet, and the first hour or two every day are usually spent in intensive fruit collection before the more laborious business of insect and spider hunting takes over. Animal prey are widely dispersed in the environment, so the group spreads out, searching mainly in the middle layers of the forest where there is most foliage and vine growth, but also investigating likely spots anywhere from ground level to the tops of trees. Some seeds and leaves are taken, and most liquid is obtained in the food itself. When that is not enough, the adaptable squirrel monkeys find water trapped in tree holes or descend to drink from streams or puddles.

The amount of forest that each group needs to utilize depends mainly on the number of individuals involved, the productivity of the forest and the amount of competition the group has in harvesting the produce. Under the right circumstances, 20 squirrel monkeys might use only 15 hectares (37 acres) whereas really big groups can use more than 130 hectares (321 acres). They do not compete with other squirrel monkey groups by trying to defend exclusive territories; rather it seems that the different groups simply avoid each other. They do, however, sometimes team up temporarily with capuchin monkeys and forage together, but such associations are not stable over more than a few hours, or a few days at the most.

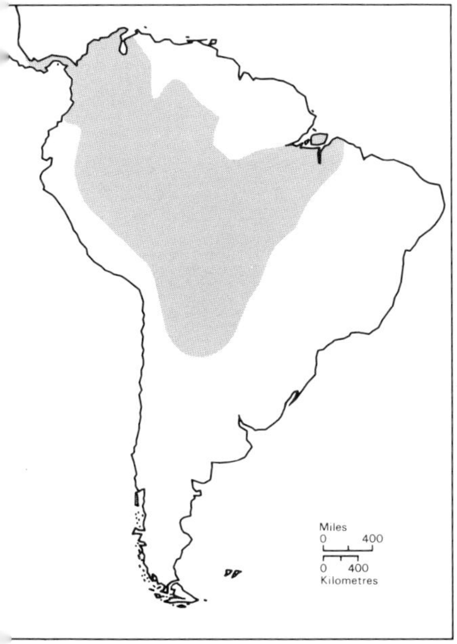

The night monkeys GENUS *Aotus*

The world's only nocturnal monkey is also the New World's only nocturnal primate (since there are no prosimians in the Americas). As with the squirrel monkey, it is not clear whether the members of this genus should be regarded as belonging to one or several species, but, either way, *Aotus* is hugely successful. It inhabits a variety of forest types from sea level to 3200m (10,500ft) in altitude, and may be found from Central America to the Gran Chaco region of northern Argentina.

No doubt, its success owes much to the lack of competition from other nocturnal primates; and it makes use of all the above-ground layers of the forest, in contrast with the West African prosimians, which have a tendency to divide up the vertical layers according to the specialization of each species. It is not, however, believed to descend to the ground itself other than on rare occasions.

Night monkeys are also known as douroucoulis (a South American Indian name that has found its way into the English language) or owl monkeys, because of their supposedly owl-like faces. Their big eyes are typical of nocturnal primates but, like the tarsiers, they betray a diurnal ancestry by the possession of a fovea in the retina: something they ceased to have any use for when they gave up being active by day. Their closest relatives are the other cebid monkeys, and the many behavioural similarities between the night monkeys and the nocturnal prosimians are the result of evolutionary convergence: over a very long time, unlike forms have become more alike as a result of adaptation to similar circumstances. Natural selection has gradually modified what was once a diurnal monkey so that, step-by-step, it became an efficient nocturnal animal. The pelage colour is a good example of this, being rather dull on most of the body but having quite definite light-and-dark patterning on the face. The dull body colour aids concealment in poor light and the face pattern can be used for signalling between members of the same species.

Night monkeys are also rather prosimian-like in the extent to which they use smells to signal to each other. Each has a gland at the base of its tail from which a brown oily substance is exuded. No doubt this substance could be rubbed straight onto objects in the environment, but the night monkey is a sophisticated scent-marker and carries its own applicator in the form of a little brush of hairs under the basal part of the tail. In all likelihood, its odours are equally sophisticated in the messages that they permit to be left in the trees. Like the lorisoids, night monkeys also urine-wash their hands and feet and print out more messages as they walk.

Scent-marking probably functions, among other things, to remind neighbours that territories are occupied; night monkeys live in monogamous family groups and seem to be quite aggressive in defending their land. Exactly how much land is defended is not known, but the territories are probably very small. Different families' sleeping trees have been found spaced out by as little as 130 m (142 yd). These sleeping trees usually have hollows into which the entire family of three or four (or even up to six) individuals may retreat by day. When that is not the case, very dense tangles of vegetation are used.

The family usually gets up at around sunset and huddles together on a nearby branch until they feel ready to face the rigours of the night. Once the mental cobwebs have been shaken off, they spend most of the night steadily searching for the fruits that constitute most of their diet, and the young shoots and leaves, flowers, sap and animal matter that make up the rest. Although they are not averse to taking a great variety of live prey, from invertebrates to birds, frogs and even bats, they seem to be rather more vegetarian than the African nocturnal prosimians. They usually have a rest at midnight, much as many diurnal primates take noontime naps, and then they forage again until dawn. They are good climbers and leapers, but definitely not in the same class as the bushbabies.

Neither grooming nor aggressive behaviour takes up very much of their time but they are quite vocal, with a repertoire that includes resonating grunts, purrs and long hoots. In spite of spending their

Night monkeys live in small family groups. The family consists of the adult pair with two offspring of different ages.

lives in the dark, they have a considerable body language; aggression is signalled by the hair standing on end, arching of the back, stiff-legged jumping, urination and defecation. It may be because these signals require to be seen to be effective that most altercations take place when the moon is at its brightest, although they are also heard on dark nights.

The males and females are very alike, as is usual for a monogamous species. Body weights vary between about 800 and 1250 g (1.75 to 2.75 lb) and the night monkeys from colder, montane forests tend to be rather shaggier in appearance than those from warmer places.

Night monkeys can tolerate freezing temperatures better than most primates but, in Argentina, extreme cold at certain times of the year sometimes forces them to huddle together all night for warmth and forage by day: a remarkable example of adaptability for a creature that is so well-tuned to a nocturnal way of life.

The titis GENUS *Callicebus*

In some parts of the great forests of the major South American river basins, the ground is either absolutely sodden for much of the year or subject to periodic flooding. In these places, the vegetation is often made up of low stature, thin trees and creepers, packed together so densely as to leave little room for the passage of any large animal. Amid such tangles, pairs of dusky titis stake out territories that may be as small as half a hectare (just over 1 acre), within which they and their immature offspring can find all that they need for their existence.

All rain forests are mosaics of different types of vegetation and the dusky titis' waterlogged haven is interspersed in many places with slightly higher ground where white-sand, nutrient-poor soils predominate, and the dense undergrowth is lacking. This is the home of the widow titi, families of which live in territories of 3 to 29 hectares (7.4 to 72 acres).

These two species are not always found together, but where they do coexist in a patchwork forest they provide a fine example of the way in which closely related species may reach a balance because of their different specialities. The dusky titi is slightly the smaller of the two weighing about 1 kg (2.2 lb) when adult, as compared with about 1.1 kg (2.4 lb) for an adult widow titi.

The ecological differences between the two species are relatively subtle, but it is clear that each is efficient enough at exploiting its preferred habitat to exclude the other, since each will encroach on the other's niche in forests that contain only one species. For example, widow titis are much more likely to be found in swamp or riverine forest when duskies are absent. This phenomenon is what biologists mean by 'competitive exclusion': a species can exploit a particular habitat but not when it has to compete with a certain other species.

Both duskies and widows concentrate on fruit for about three-quarters of their diet, but whereas duskies make up much of the remainder with foliage, widows do so with invertebrates. This difference affects the way in which they organize their daily time-tables: both feed intensively in the early morning, but in the middle of the day the duskies take a break while the widows forage for insects. This is quite understandable given that leaves are relatively easy to collect but take time to digest, whereas insects and spiders have to be searched for individually yet are relatively easy to digest. Furthermore, a specific quantity of invertebrates gives the eater very much more energy which can be used for physical activity than does a similar quantity of foliage. Presumably, the duskies are busy digesting leaves at noon. Even the widows eat some leaves before settling down for the night, which they do as much as two hours before dusk. The duskies go on feeding later, so that both species end up having been active for about the same number of hours, although different shift systems have been employed.

They also have rather different preferences in sleeping sites. The duskies choose dense tangles of vegetation in the middle layers of the forest, but the widows tend to select large branches in emergent trees just above the main canopy level. Thus, the duskies sleep where they can neither see well nor be seen, whereas the widows have a high vantage point.

Both species advertise their territorial ownership with loud, resonating calls that are usually given at dawn; and, in this, the widows have the advantage of not having to get out of bed to be in the best place to hear and to transmit the low-pitched sounds. Vegetation is a great acoustic absorber, so it is better to be above it during a shouting match. Of course, it is unlikely that the two species are actually calling to each other, even though they may often hear each other. Territorial calls typically function to maintain the spacing between different groups or families of the same species, since it is individuals of these social units that are most in competition with each other. A dusky titi competes much more with another dusky titi for everything it requires than it competes with a widow titi that has slightly different needs, even if the two of them share some parts of the forest. It is a fallacy to suppose, as many do, that animals arrange their lives 'for the good of the species'.

Titis protect their interests for the good of their immediate families. Their monogamous life style epitomizes the possible perfection of that system; the general tenor of their lives is overwhelmingly amicable. Careful grooming of each other's fur is an important activity between all members of the family; and when they sit or sleep together, they intertwine their tails so that sometimes three or even four tails are wound around each other, hanging beneath a branch.

Male titis are possibly the most concerned of all primate fathers, being responsible for all of the carrying of their babies from the second day, except when it is time for their mothers to feed them. The young titis rapidly learn to climb from one parent to the other at the appropriate time. It even seems likely that, when the time comes for them to leave the family and strike out on their own at the age of two or three, they do so without any aggressive parental hints.

A third species, the masked titi, inhabits the Atlantic coast region of Brazil. It is intermediate between the other two, both in size and in habitat preferences, being less inclined to specialize in a particular part of the forest. In the absence of any competing titi species, it has been able to maintain a more generalist approach to life.

For illustration of a dusky titi (see page 94).

The dusky titi is the smallest member of its genus. There are seven races inhabiting some of the wettest forests of South America.

The howler monkeys GENUS *Alouatta*

In the early morning, as the mist still clings to the damp forest trees, loud grumbling roars resonate through the still air. At first, they come from one direction, then another, and another. These deep and impressive calls are the 'howls' of the New World's loudest primate genus. Six species of howler monkey are found between northern Argentina and the northern border of the humid tropics in Mexico. All of them have greatly enlarged lower jaws that accommodate egg-shaped resonating chambers and permit the monkeys to make their unique, reverberating calls.

These chambers are bigger in the males than in the females, and in consequence the male calls are much louder and deeper. All howler monkeys show considerable sexual dimorphism, with the females weighing only about three-quarters as much as the males. They are not the biggest of the American monkeys, but males weigh about 7.5 kg (16.5 lb) and females about 5.7 kg (12.6 lb). They vary in colour from spectacular reds, through various shades of brown to black, and the so-called black howler from the south of the Amazon basin is one of those rare mammals in which the males and females

RIGHT **The adult male red howler monkey makes a loud roar, one of the most spectacular sounds of the South American forest.**

BELOW **This species is known as the black howler monkey, but it is only the male that is black. This is a normal coloured female.**

are quite differently coloured. In this case, it is the males that are black whereas the immatures and the adult females are pale grey.

Among monkeys, truly prehensile tails are found only in New World species, and in only a few of these. The howler's tail ranks somewhere between that of the capuchin and the spider monkey in its prehensile dexterity, and it has a flat 'palm' of bare skin on its underside for about a quarter of its length from the tip. Howlers use their tails almost constantly, anchoring themselves with them when they are asleep, and either holding them at the ready or actually looping them around any convenient support while awake. The tail seems to be important as a safety line, since a feeding howler will always use it to grasp a branch, even if its weight is fully supported by its arms, legs or bottom. Sometimes, it hangs entirely by its tail but it does not seem to do this as readily as a spider monkey may and any nearby twigs are often clasped with the feet at such times, perhaps more as psychological than as real support. Howlers also use their tails to ward off insects when they are trying to rest, and both sexes sometimes manipulate their genitals with the tips.

They are mostly rather slow and deliberate in the way they move about in the trees, although once in a while they are motivated to make quick rushes along branches. Both their hands and feet are adapted to give a powerful grip and they grasp lianas and branches with a steady ease. They seldom descend to the ground, and when this does happen they are awkward in their movements, for it is difficult for them to place their hands and feet comfortably on a flat surface.

The steady nature of a howler's progression in the trees is appropriate for a monkey that consumes leaves and flowers for about half of its diet. Such material is low in energy and takes time to digest. Unlike some of the Malagasy prosimians and the colobine monkeys of the Old World, howlers do not have digestive tracts especially adapted to cope with vast quantities of fibrous material, so they apparently have to be extremely careful to select the young and tender, and therefore more digestible, foliage.

Young leaves generally contain more protein and soluble sugars than mature ones. The remainder of their diet is comprised of fruit and any maggots they find therein. Presumably, they are unable to depend on a more fruit-based diet because the supply is less reliable than that of the almost ever-present young foliage. Howlers are the most folivorous of New World monkeys and their ability to subsist on foliage when necessary gives them an advantage over animals that cannot do this, whenever fruit is in short supply.

Howlers live in groups of about 8 to 20 individuals, with two to four times the number of adult females to adult males. They move slowly about their home ranges and may not cover the entire area in a month. Home ranges vary greatly in their size, depending on the local conditions, but 500 m (550 yd) would probably be a reasonable estimate of the distance that a group might travel in one day.

They are chiefly inhabitants of tall forests containing a large proportion of mature trees which have not been disturbed by man, but they may also be found in less virgin areas, including relict patches of forest left when much of the natural vegetation has been destroyed. The way in which the groups divide up their habitat depends very much on the population density and the food supply. When few groups occupy a relatively large area with plenty of food, they tend to space themselves out into exclusive territories; but when conditions are less ideal, the home ranges overlap, although the groups continue to avoid each other for most of the time. Whatever the spacing between the groups, it is maintained first by the dawn chorus of howling that indicates the location of each one to the others, and secondly by the howls and grunts that opposing troops direct at each other when they meet. These vocal competitions substitute for real fighting to the advantage of all concerned.

Although howlers live in fairly permanent, cohesive groups, within which the social relationships seem to be acted out in a distinctly low-key fashion, it seems that they have adopted a simple existence in comparison with, say, squirrel monkeys or macaques. They certainly have relatively small brains for their body size, something that could be a consequence of little pressure to show any social intelligence. They do not even groom each other's fur, which is unfortunate as many suffer infestations of the larvae of the botfly. The eggs of these rather nasty grubs are laid in their fur and, after hatching, they burrow through their host's skin into its flesh. Depending upon the number of botfly larvae a howler is carrying, the consequence could be discomfort, debilitation or even death.

Howlers in some areas also suffer from periodic epidemics of yellow fever that wipe out large numbers of monkeys. On Barro Colorado island in Panama, the local population of mantled howlers used to grow and crash in a regular seven-year cycle until forest destruction broke the connection between the island monkeys and the source of the disease in eastern Panama. Such massive changes in the population must have a turbulent effect on the social organization, wiping out certain classes of monkeys within the group – all the adult males might be eliminated for example – if not certain groups in their entirety. This, in turn, is bound to affect the pattern of home range use and the extent to which individuals transfer between groups.

On a more optimistic note, howlers do not seem to be bothered by birds of prey. Probably, the adults are too big to be taken, although the infants and juveniles might be expected to fall prey to eagles. Howlers certainly fall prey to humans in many parts of South and

Central America because they have enough meat on them to justify the effort of hunting and, if necessary, the expense of a cartridge. In addition, by being slow moving, they make themselves easy to shoot. At the present time, the brown howler of eastern Brazil and the Guatemalan howler of Central America are threatened by a combination of forest destruction and hunting for the pot.

The capuchins GENUS *Cebus*

The capuchins are the hustlers of the American primate world. When they go through a patch of forest in search of food, they really do a thorough job. They spread out through all the different levels, from the low understorey to the tops of the tallest trees, and systematically search for anything that is edible. Fruit is the main item in their diet, but they also eat flowers and young shoots, and take large numbers of insects and spiders. They search for live prey by ripping off loose bark, turning over and unrolling leaves, pushing aside logs, splitting hollow vines and searching among the debris that accumulates at the bases of palm fronds. Along the way, they consume eggs from nests, nestling birds and squirrels, lizards and may even kill and eat smaller monkeys. In Peru, a small group of tufted capuchins has been seen to prey upon a dusky titi, and white-faced capuchins are rumoured to attack squirrel monkeys in Panama.

Capuchins are among the most intelligent of all monkeys and can adapt quickly to their local conditions. As evidence of this, tufted capuchins are the only monkeys to gain access to the food inside some of the harder shelled nuts in the Amazon forests. They do so by hitting them repeatedly against the trees or by smashing two nuts together. White-faced capuchins harvest oysters from the seashore at low tide; and in many places capuchins have turned their talents to crop raiding. Peasant farmers can find it a challenge to harvest their corn before the monkeys reach it.

Because of their versatility, capuchins may be found in almost every type of forest somewhere in Latin America. Of the four species, only the white-faced is found in Central America, its geographical range stretching from Belize to northern Colombia. The most widespread of the other three species is also the most widespread monkey in the New World: the tufted capuchin is found from Colombia to northern Argentina and the Atlantic coastal forests of Brazil.

Wherever they are found, capuchins play an important role in the ecology of the forest. They pollinate flowers by moving from bloom to bloom; and it has been estimated that one group of white-faced capuchins could eat and later disperse more than 300,000 tiny seeds of a particular tree species in a single day! Take away the

ABOVE **The tufted capuchin has the widest geographical range of any New World primate.**

RIGHT **Fine eye-hand co-ordination allows this young male white-fronted capuchin to pick up even the tiniest of objects with great precision.**

capuchins, and the replacement of each generation of trees by the next would be bound to suffer. The importance of monkeys in seed dispersal is well known and some seeds are significantly more likely to germinate if they are passed through a primate's digestive system. Others simply germinate quicker with the monkey's help than is the case if they simply fall from the tree or are carried off and dropped without being eaten. In countries where the local people prize forest fruit, they may be benefiting more from the presence of the monkeys than they realize.

Capuchins live in groups that vary in size from about 6 to 30 individuals. Almost always, there are more females than males among the adults of any one group. At night, little clusters of two, three of four monkeys huddle together in the forks of tall trees. By day the group spreads out so that 30 monkeys can easily be dispersed over several hundred metres. The amount of forest that a group uses is highly variable and depends very much on the amount of food available. Similarly, some capuchin groups defend their territory by calling at others and chasing and fighting with rivals at the boundary, whereas in other areas there is no sign of territorial defence.

There is considerable variation in body weight in the genus, and adult males, which average about 3.3 kg (7.25 lb), weigh rather more than the females – about 2.6 kg (5.7 lb). Any capuchin, however, would seem to be small enough to be taken by quite a variety of predators; and the monkeys show signs of alarm to such potentially threatening creatures as birds of prey, ocelots, boa constrictors and even caimans. The last is a member of the alligator family and may catch the occasional monkey that is a little too casual when it descends to drink from a river or pond. Capuchins drink every day, and although they will often use water that collects in tree holes, that is not always available.

Since capuchins are such sharp-witted animals, they are probably not caught very often. A predator would have to avoid being seen by any member of the group, since the first one to see him would warn the others. They have a very large communicative repertoire and signal to each other by means of body movements, facial expressions, calls, grinding their teeth, smells and touching, hitting, biting and climbing onto each other. For example, a young male might indicate by stretching out on his back that he would not mind if another monkey came and groomed him; or he might threaten others by walking stiff-legged with his back slightly arched, a little like that of an aroused cat.

The highly mobile lips, tongue and eyebrows, and also the teeth and ears, are all brought into play for the facial expressions. Males and females of the white-faced capuchin have the rather touching habit of puckering their lips at each other, as if blowing kisses, before and during copulation. Like all other monkeys, they copulate with the female below the male and facing away from him, as is most often the case among mammals.

When an adult male is approaching an attractive female, he may indicate his intention by giving a high-pitched warble; this sometimes seems to have the effect of persuading the female to accept him. The rest of the calls of capuchin monkeys consist of a complex variety of barks, growls, loud 'caws', screams, chatters, trills, twitters, purrs, whistles and so on. Many of them seem to have fairly specific meanings, but the contexts in which they are given may also be important to a listener's interpretation of the meaning. Once again, the complexity of the repertoire is evidence of the capuchin's high intelligence.

There is one way in which these monkeys are rather less talented than some of their larger American cousins and that is in the use of their tails. A capuchin does have a prehensile tail that he uses to hang onto branches and tree trunks, and he can hang by his tail when feeding or playing; but it lacks the 'palm' of bare skin found on the tails of the larger, long-tailed genera. If this is in any way a deficiency, capuchins certainly make up for it by having the most dextrous hands of any New World monkey. Their slender fingers are ideal for manipulating small objects or grooming each other's fur, and they can grasp things securely between the thumb and the fingers. They are a fine example of a monkey that combines intelligence with skilled, coordinated use of its hands and eyes.

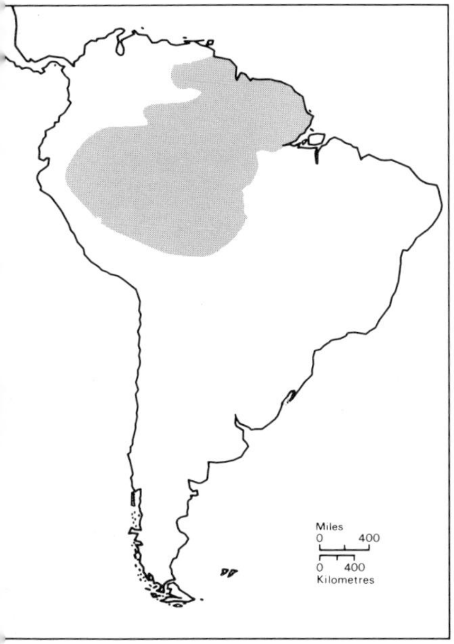

The sakis GENUS *Pithecia*

In Colombia, they are known as *monos voladores* (flying monkeys). In Guyana, they have names like flying jack and breezy monkey. Wherever they are found, all four species of saki are noted for their spectacular locomotion. Like most other primates, they can walk slowly on all fours along horizontal branches; but when the time comes to move a little quicker, they bound and gallop through the trees, hop bipedally like arboreal kangaroos and leap downwards across big gaps. A distance of 10 m (33 ft) between supports presents no great problem to a suitably motivated saki. About the only ways of moving in the trees that it will not use are to swing under branches like a gibbon or to hang by its tail. But then, with its existing repertoire, it hardly needs to.

The sakis are quite closely related to the bearded sakis and the uakaris: together they comprise the pithecine subfamily. Sakis are small monkeys, weighing between 1.4 and 2.5 kg (3 to 5.5 lb), and the females are a little smaller than the males. They all have rather coarse, thick coats and big, fat-looking, non-prehensile tails. Their

LEFT The shaggy coat and thick tail are typical of the monk saki, one of the best leapers among the South American monkeys.

BELOW The monogamous pair is the basis of society for all sakis. Here, a female white-faced saki grooms her mate.

faces tend to have a rather flat appearance, dominated by an extraordinarily broad nose. Sexual dichromatism crops up again in this genus: male white-faced sakis have black body hair and a broad white ring of fur around the face, whereas the females have a brownish pepper-and-salt appearance, with rather pale chests and bellies and a white stripe down either side of the nose. Rather surprisingly, female white-faced sakis look much more like both sexes of shaggy saki than like their own mates. The males develop their black-and-white coloration during the first three months of their lives.

Sakis are found in various types of forest, including, on occasion, areas of quite new growth and young trees, but they clearly do much better in older, undisturbed habitats. In addition, they do not seem to like forests that are regularly inundated by flooding.

They live in small family groups, consisting of an adult pair and their immature offspring but can sometimes be seen in larger aggregations. Only one baby is born at a time, and the parents seldom seem to be accompanied by more than one young. In contrast to other monogamous South American primates, almost all of the infant care seems to be given by the mother, but the babies grow quite rapidly and by about four months they start showing signs of moving about on their own and stealing solid food from their mothers. The natural history of the genus is not well known but it is possible that the young sakis strike out independently before they are a year old. If so, this is a very rapid physical and social development for a monkey.

Sakis seldom, if ever, descend to the ground or climb to the tops of the tallest trees that emerge above the main canopy. Both places probably hold too much danger for them, either from small cats, big snakes and other carnivorous animals on the ground, or from birds of prey up above. The middle and lower levels of the canopy and any understorey below that offer sakis their best available food and protection. When they are alarmed, they alternate between fleeing rapidly and freezing, which can be very confusing for any animal that tries to pursue them. The technique is rather reminiscent of that employed by pygmy marmosets.

Fruits and seeds seem to form the basis of their diet, with considerable variations between seasons and between the sakis of different localities. Palm fruits are certainly popular, and flowers and young leaves and shoots are taken occasionally. Insects seem to be eaten only very rarely. Whatever they feed on, saki families seem to meet their needs from relatively small areas of forest in home ranges that overlap with those of their neighbours. In a forest near the Brazilian city of Manaus, one family spent most of its time in an area of 2.5 hectares (6.2 acres) and never ventured out of a total area of 9 hectares (22 acres).

In contrast to the closely related bearded sakis, monkeys of the genus *Pithecia* have opted for a rather unobtrusive way of life among the tree trunks and creepers of the lower forest levels, where they leap from one support to another and manage to subsist on small sources of food by living in very small family groups.

The bearded sakis GENUS *Chiropotes*

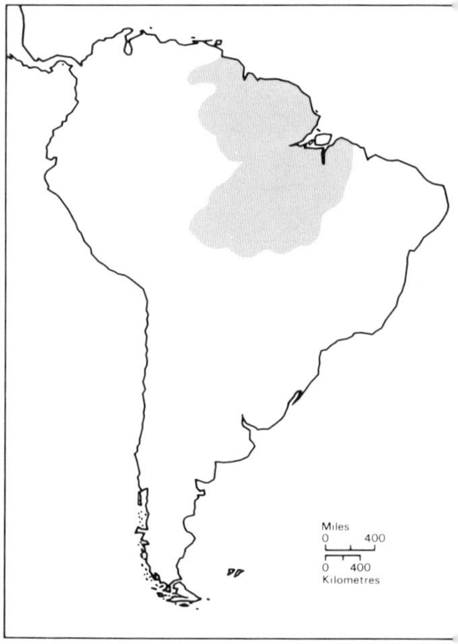

Back in the days when it took months to travel between the urban natural habitats of scientists and the rather greener natural habitats of monkeys, taxonomists often gave new species their names on the basis of whatever preserved skins and bones were available. Live specimens were often unknown in captivity, and the namers obviously did not have the benefit of colour photographs to show what the animals really looked like. If they were lucky, they might have an artist's impression from which to work.

One of the two bearded saki species is a classic example of what can happen when the preserved specimen no longer looks like the original creature. The specific name of *Chiropotes albinasus* is derived from two roots: *albi* meaning 'white' and *nasus* for 'nose'; hence it is, or ought to be, the 'white-nosed saki'. (Bearded is dropped as being too much of a mouthful.) Unfortunately for the taxonomists who named it in 1848, the animal concerned has a bright red nose when it is alive, a nose that fades to white in a preserved skin. Nevertheless, the name 'white-nosed saki' has persisted, and it is only recently that primatologists from its native Brazil have started to try to popularize 'red-nosed saki' as its English name. Although there are no formal rules governing non-scientific names, it helps if they make some sort of sense and are widely accepted.

The second bearded saki species has two quite well-defined races that are collectively known in English simply as the 'bearded saki'. The lack of any more sophisticated English names for monkeys of this genus is a fair reflection of the paucity of our knowledge about them. Although recent studies have improved the picture somewhat, they remain among the least known of South American monkeys.

All bearded sakis are predominantly black, with long, thick tails and rather short body hair. Both sexes have bright pink private parts which, of course, show up very well against a black background. Both sexes also have the beards from which their name is derived, and the males have extraordinary, bulbous swellings on the tops of their heads.

They are very definitely creatures of tall, virgin forest, although they seem to be able to tolerate different types, from swampy places to drier areas, some of which are at quite high elevations. Whatever the type of forest in which they are found, they avoid the lower levels of the vegetation and spend most of their time in the upper part of the canopy.

Adult bearded sakis weigh between 2.6 and 3.2 kg (5.7 to 7 lb), with the females being perhaps fractionally smaller than the males: the number of individuals that have been weighed is too small to be more exact. They live in groups of up to 30 or more monkeys, with several adults of either sex coexisting within each social unit. Such a group can easily cover 3 km (nearly two miles) in a single day's travel, and they probably have home ranges of 2 to 3 km^2 (0.8 to 1.2 square miles)

The bearded saki's whole way of life is based upon the consumption of a large amount of high-energy food and the expenditure of that energy in considerable activity. They move very long distances between the trees that yield the best seeds and fruit, since these make up most of their diet, and they feed very intensively when they get there. They also take flowers and occasional leaves, but it is their specialization in seeds that is really characteristic. They can eat numerous unripe seeds that other monkeys avoid.

One of the ways in which many plants defend their seeds against such depredations is by encasing them in hard shells, as in the familiar Brazil nut. The dentition of the bearded saki is adapted to get over this problem; the incisors (front teeth) project forward, with the upper set positioned to overbite the lower ones like the arrangement of a parrot's beak. The teeth are extremely strong and capable of breaking open very tough husks. Similar dental adaptations are shown by the other genera of the subfamily Pitheciinae, the sakis and the uakaris.

Because they can eat seeds that are in various stages of ripening, bearded sakis can find large amounts of food in single trees and are thus able to forage in large groups. However, it seems that these groups may consist of a number of family units – mated pairs and their offspring – that do not live together permanently. Individuals appear to be quite friendly with each other; at least they indulge in fairly frequent mutual grooming of their fur. They also have a long whistle which they use for communication within the group when they are spread out among several trees. In this way, they do not lose contact with each other. During the wetter months of the year, they are commonly found in association with capuchins and woolly monkeys.

Bearded sakis are much prized by the local people both as food because of the quality of their flesh, and because their tails are traditionally used as dusters. They are therefore subject to regular hunting pressure; and in recent years, some populations have become very severely endangered by the destruction of the forest to make way for cattle-ranching projects.

For illustration of a white-nosed saki (see page 106).

LEFT This monkey was named the white-nosed saki on the basis of a skin that had dried and faded. Red-nosed saki is obviously a better name.

BELOW All uakaris seem to prefer forests that are often flooded. This is one of the white race of bald uakari.

The uakaris GENUS *Cacajao*

All South American monkeys have long tails – with one exception: the uakaris. These have tails that are rather short and bushy. The monkeys themselves have a generally rather shaggy appearance, which they can accentuate by making their hair stand on end when they are excited. This also makes them look much bigger than they really are. The males, at about 4 kg (8.8 lb), are in reality rather bulkier than the females, which weigh about 3.5 kg (7.7 lb), and they may be further distinguished by their longer canine teeth and prominent bulges of jaw muscle attachment on top of their heads.

Uakaris (pronounce the 'u' as a 'w') are found only in the forests of the Amazon basin, where they clearly prefer the sorts of forest that are regularly inundated with floodwater. Indeed, they may be entirely restricted to these wet areas.

There are two species: a little-known black one, and a better-known (at least in zoos) bald-headed one that has red-haired and white-haired races. The red uakari, with its crimson face and variety of gruff or outraged expressions, reminds many South Americans

RIGHT **The red race of the bald uakari reminds Brazilians irresistibly of gin-drinking Englishmen.**

irresistibly of the sort of British expatriate who indulges in too much gin – and it is therefore known locally as 'the English monkey'.

Although bald uakaris are not uncommon in captivity, they have yet to be studied properly in the wild. Fruit is probably the most important item in their diet, although they certainly eat quite a few leaves and flowers, and may well take such live prey as snails and insects.

Both species live in groups containing several adults of either sex, as well as their young. There are reports of groups of up to 100 individuals, but it seems more likely that totals of 15 to 30 are more realistic. Social relations seem to be fairly amiable, with little fighting between adults and with the usual primate playfulness among the young. Uakaris seem to groom each other more than any other New World monkeys, and this may be important for the health of their long pelage as well as for strengthening the friendly bonds between individuals. A uakari that is being groomed typically luxuriates in the attention it is receiving until the groomer has had enough and stops, at which point the roles will probably be reversed.

Newborn infants lack the generous body hair of their parents and are carried between the mother's hip and her body, where they benefit from her warmth. By the time they are about three months old, they have their shaggy coats and they begin to spend more time on their mothers' backs. Young bald uakaris also grow a thatch of thick, grey hair on the head, something that is not lost until they are approaching adulthood.

The time it takes for a uakari to grow up seems to be surprisingly variable, at least among the males. Females may become pregnant not long after their third birthdays, but males may enter a subadult phase at this time. If so, they can spend the next two or three years rather resembling adult females in a number of ways: they lack adult male muscle bulges on their heads, remain at about the size of adult females, call in response to minor disturbances (which adult males do not do) and have a rather female appearance to their genitals, with the small, folded scrotal sac being reminiscent of the female labia.

When a male enters adulthood, it rapidly puts on muscle and the testes descend into its scrotum. It also changes its behaviour, becoming less interested in trivial disturbances but much more ready to respond aggressively to anything that appears to present a serious threat to the group.

The surprising thing is that male uakaris can skip the subadult phase altogether and reach adulthood when only about three years old. It seems to be possible that social factors determine whether or not this occurs and that male uakaris have evolved the ability to put off growing up until conditions favour their success as adults. This is speculation, but a highly sophisticated response by the body's hormones to factors in the social environment would seem to be involved. It is, however, also possible that different rates of development of different individuals are more rigidly pre-programmed, but either way the phenomenon is most unusual for a monkey.

The woolly monkeys GENUS *Lagothrix*

To many people, Humboldt's woolly monkeys are familiar, rather attractive, chunky monkeys covered in dense, short hair and with long, prehensile tails. They may be brown, grey or nearly black, and the infants have an appeal that has made them popular as pets or in the children's sections of zoos. Their winsome looks are somewhat misleading, for the adults, particularly the males, are well-armed with long canine teeth, which they are not afraid to use. They can also grow to a considerable weight, 7 kg (15.4 lb) being about average. Females are usually about 85 per cent of the male weight but have longer tails.

The approach of a group of these animals through the forest can be a spectacular event, and not a little awe-inspiring the first time it is seen. Loud rustles and crashes in the trees precede glimpses of dark bodies running along branches or swinging underneath by the arms and tail. Springy branches fly upwards as strong hands release them and, if the monkeys are unafraid, they may approach and gaze at the human visitor, while suspended just a few metres above his head. They look much bigger at this angle!

Sadly, they have been persecuted by man in many places because they yield meat for hungry bellies and babies for the pet trade. In respect of the latter, they are an example of a monkey that is too attractive for its own good, since the life expectancy of a pet is short, even if it survives intact the fall when its mother is shot.

Baby woolly monkeys have lighter coats than their parents and are usually carried on their mothers' bellies. When they get a little older, their coats darken and they spend more time on their mothers' backs, although they are capable of hanging onto almost any part of the females' bodies. They show no signs of discomfort even if the females move rapidly in the trees or indulge in vigorous play. For a woolly monkey, playing involves wrestling, hanging by the tail or other limbs, chasing and jumping onto each other and swatting at leaves with their hands. Participants usually open their mouths and bare their teeth, presumably as a signal that all the rough stuff is only meant in fun. The big males tend to be a bit stand-offish at these times, but the adult females quite often join in. Not surprisingly, it is the youngsters that play most frequently, and it is only they that play by themselves.

A lot of the playing takes place in the middle of the day when the

The woolly monkey's prehensile tail has a sensitive pad at the end like the palm of a hand. For the monkey, the tail is a versatile and muscular fifth limb.

group stops foraging or travelling and the adults spend much of their time dozing. This is also the time when monkeys groom each other, working carefully through the dense fur with their fingers and teeth.

Woollies live in groups of about a dozen individuals, in which there are several adults of either sex. They live only in relatively undisturbed forests and do not seem to defend territories, but the amount of land over which they roam is unknown. They eat various kinds of leaves, fruits and flowers, and chew bark and the wood of small twigs. Dead leaves are sometimes unrolled and inspected, but woolly monkeys do not systematically inspect one dead leaf after another as capuchins will do in the same patch of forest. Insects seem to be a very minor part of a woolly monkey's diet, if they are included at all.

Two animals will sometimes feed amicably on the same bunch of leaves, which is in keeping with the generally affectionate tenor of their lives. A greeting between two woollies consists of a brief mouth-to-mouth kiss, followed by trembling of the lips at each other. Sometimes they also nuzzle each other's necks or embrace each other. Overt aggression is rare within the group.

There is a second species, the yellow-tailed woolly monkey, which is so rare and inhabits such inaccessible montane forests that it was thought to be extinct until it was rediscovered in the High Andes of Peru in 1974. It is found in dense, foggy forests at an altitude of 1800 m (5900 ft) or more above sea level; but the first live animal actually seen by scientists was a juvenile that a soldier was keeping as a pet when biologists who were scouring the Andes for a sign of the species visited his home. Since then, it has been seen in the wild at more than one location.

RIGHT It is not difficult to see how the spider monkeys got their name. This species is the black spider monkey (see page 112).

BELOW When the branches are too far apart for her infant to climb across by itself, the adult female Geoffroy's spider monkey makes a bridge out of her body (see page 112).

The spider monkeys GENUS *Ateles*

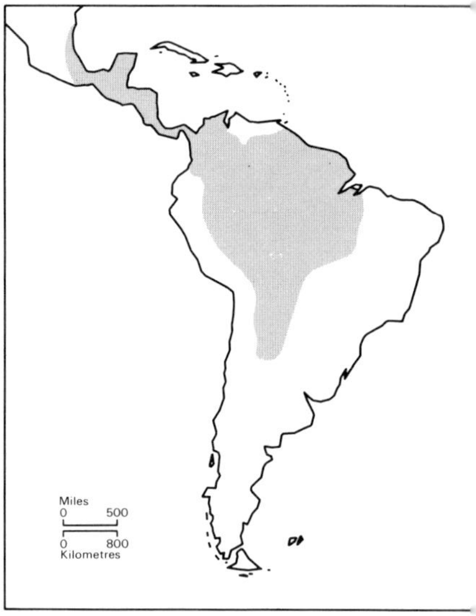

In the Amazon rain forest, some fruits are found growing abundantly on large trees, from which they may be harvested in great quantities, and others are found a few at a time on relatively small trees. At certain times of the year the former are most easily found, but at other times fruit-eating animals must forage for the latter.

Spider monkeys depend for the vast majority of their diet on ripe fruit and they have evolved an extremely flexible society that permits them to capitalize most efficiently on the changing nature of the supply. They move around the forest either singly or in aggregations of anything up to about 20 individuals. These aggregations are not permanent units; they may last for a few hours, or sometimes for a few days, as animals either join or leave, either individually or in still smaller aggregations.

When the larger food sources at the big trees are most widely available, the biggest aggregations of monkeys are found. Typically, five or more of them move together and feed together in the big trees. It is, however, noticeable that, when they do feed in a large tree, they space themselves so that they are not in close proximity to one another. They maintain this spacing by continually adjusting their positions in relation to one another, and latecomers generally wait until earlier arrivals have left before entering the tree. It seems that spider monkeys can be quarrelsome feeders if they are too close to each other, and this spacing out saves them all trouble.

During those months of the year when they have to depend on small, scattered sources of fruit, such as from palm trees, lone individuals and smaller aggregations are found moving through the forest. Thus, they avoid quarrelling at food sources with only enough ripe fruit at any one time to feed a very few monkeys. During this time, they reduce the average length of a day's travel from almost 5 km (about 3 miles) to about 700 m (766 yd).

From the ease with which individuals move in and out of association with each other, it might seem that spider monkey society is quite fluid and unstructured. It appears, however, that within any one area the population is divided up into groups, as is usually the case with monkeys. A certain patch of forest might have two or three groups within it, all moving around as subgroups. Whenever two subgroups meet, their reaction to each other depends very much upon their membership of the permanent groups. If they belong to the same group, they may mingle and divide up later, perhaps along different lines; but if they belong to different groups, no such thing will occur, Then, when the males are within about 100 m of each other, they will mutually threaten with a great deal of bluster, including chasing about in the trees, shaking branches and whooping and growling at each other. Sometimes a male will

For illustrations of a spider monkey and an adult female Geoffroy's spider monkey and infant (see page 111).

also smear saliva and a secretion that comes from a gland on his chest onto the branches, presumably to deposit his scent in the area. These altercations can easily last for an hour or more but they seem to be strictly male affairs; females remain quietly in the background.

The long-haired spider monkey of southern Colombia is typical of the genus in behaving like this, but it is likely that all the species and races are rather similar. It is generally accepted that there are four species of spider monkey in the forests between Mexico and the southern part of the Amazon basin, but this is largely a question of default in the absence of detailed information. Sixteen races are usually accepted but some are not well enough known for taxonomists to be entirely comfortable with the present arrangement. Possibly, fewer species should be recognized.

All are, however, rather similar in size and shape. With their gangly limbs and the most mobile and dextrous prehensile tails of any primate, spider monkeys are aptly named. They are extremely agile in the trees. They have a rather scuttling, characteristic way of running along flat surfaces, to which they add great skill at swinging beneath branches, suspended by any one or more of their five limbs. Their thumbs have been greatly reduced by natural selection for a hook-like hand that is ideal for fast swinging; either a spider monkey has no thumbs at all, or it has the merest of stumps.

Males and females are much the same size and usually weigh about 8 or 9 kg (about 17.6 to 19.8 lb), but at 13 kg (28.7 lb), one individual, a wild male black spider monkey from Brazil, is the biggest known South American primate. On average though, spider monkeys are not as large as the woolly spider monkey.

The woolly spider monkey GENUS *Brachyteles*

Like the lion tamarin, the woolly spider monkey has the misfortune to live in the rapidly disappearing forests of Brazil's heavily populated Atlantic coastal area. As its name would suggest, the single species of this genus is closely related to both the woollies and the spider monkeys. Few individuals have been weighed, but at a hefty average of 10 kg (22 lb), the woolly spider monkey can lay claim to the title of biggest South American monkey.

Like its close relatives, it has five long limbs including a highly dextrous and sensitive tail which it can use for hanging about in the trees or for picking up small objects. It is so rare that very little is known about its private life, although it is suspected of having a rather flexible social organization along spider monkey lines. Groups of about 10 to 30 individuals are believed to split up during the day as they forage for seeds, fruits, nectar, leaves and possibly insects. Then, in the late afternoon, they call loudly to each other and

The woolly spider monkey, or muriqui, is the biggest primate in the New World and one of the rarest.

reassemble at a suitable tree in which to sleep.

Probably on account of their size, they seem to be regarded as tough customers by the other local monkeys, being avoided by howlers and capuchins. Masked titis even cry out in distress at their approach, although it is not known what the woolly spider monkeys have done to intimidate them. However, they are certainly not exactly quiet or shy creatures, having a vocal repertoire that includes whines, grunts, loud belches, huffing and puffing, and loud, horselike neighing. As if that were not enough, groups tend to react to being disturbed by screaming in chorus for up to half an hour at a time.

The outlook for the woolly spider monkey is bleak, to say the least. Only a few hundred individuals still exist in the wild and it does not do well in captivity. Early in the nineteenth century, explorers' expeditions found so many of these creatures that they reported them to be a reliable source of meat. Now they are too scarce to be relied on and are, anyway, totally protected under Brazilian law; but they still end up in the cooking pot on occasion and legal protection is useless if what is left of the forest is allowed to disappear. The species has recently been adopted by conservationists as a major symbol of Brazil's fauna in the hope that a wider awareness of its plight will lead to measures to aid its salvation. If not, it may soon join the dodo as a symbol of extinction.

OPPOSITE **Guerezas are easy to spot in Uganda's Kibale Forest, but they leap away at the slightest disturbance.**

5. THE MONKEYS OF THE OLD WORLD

Throughout Africa and Asia, there is just a single family of monkeys, the Cercopithecidae. Unlike the New World monkeys, they are not restricted to tropical and subtropical forests, although that is where most of them occur. In Africa, several species have made a life for themselves in the more open country of the savannahs and montane grasslands; and, in both continents, there are monkeys that have adapted to life around human habitations, often raiding crops or rubbish dumps for a living.

The spread of the Old World monkeys has been limited by oceans, deserts and cold, temperate conditions. They have not been successful where either a lack of water or very low temperatures in the winter eliminates the presence of a year-round supply of vegetable food. No monkey has ever learned to hibernate to overcome this problem.

In China and Japan, there are species that live at the very limit of endurable conditions, reaching about 41°N. In Africa, a single species survives in the warm, temperate forests of Morocco and Algeria up to about 36°N. There are no living monkeys that naturally occur in Europe, although fossils indicate that they were once there. The southern parts of Asia and Africa present no such limiting cold climates, but no wild monkeys have managed to reach Australia or Madagascar.

Old World monkeys typically rest in a sitting position which they are able to do without, apparently, getting sore bottoms due to the presence of 'ischial callosities': tough pads that are located in just the right place. The various species are divided into two quite distinct subfamilies: the Colobinae and the Cercopithecinae, each of which is found in both Asia and Africa. Only two genera are common to both continents: the macaques, which are Asian but have one African species, and the baboons, which are entirely African except for a few members of one species that live in Arabia.

In common with the apes and ourselves, the Old World monkeys have nails, rather than claws, on all digits; and none has a truly prehensile tail. All species are diurnal and smells play only a relatively minor role in their lives in comparison with either the prosimians or the New World monkeys.

I The Leaf-eaters of Africa and Asia: potbellies and odd noses

The colobine monkeys are generally referred to as being leaf-eaters, or sometimes just 'leaf-monkeys', because they all have guts that can process relatively large amounts of foliage. All Old World monkeys rely mainly on plant products for their diets, but the colobines have pursued a specialized course in their evolution, a course that has definite advantages but which is not without its costs.

In order to survive, they need to consume regular amounts of protein, carbohydrates (sugars) and other nutrients, as do all primates. It is important that the supply should be available all the year round, since no monkey can hibernate. Fruits are an obvious potential source but they are not always available in sufficient quantities, and although they contain plenty of energy-giving sugars and oils, they are generally rather low on protein. Seeds and flowers are sometimes a good source of protein, but they tend to be both rather variable in quality and only sporadically available.

The best source of protein in the rain forest is the foliage that is all around the monkeys for twelve months of the year. Furthermore, leaves also contain large amounts of carbohydrate in the form of cellulose and related compounds, in their cell walls. Thus, it would seem to be a good idea to eat large quantities of them. Unfortunately, it is not as simple as that. In the first place, young leaves usually contain very much more protein than mature ones, so it pays to be selective. Of course, leaves are not young for very long, so young ones take a little more looking for than do older ones.

In the second place, no mammal can digest plant cellulose by itself. They all need the help of microbes in the gut (gut microflora) if they are to do so. And although probably all mammals, including man, harbour some gut microflora capable of digesting cellulose, the process requires the food to be passed along very slowly, and the animal to be able to absorb the products after the flora have broken down the cellulose.

The more folivorous of the Malagasy lemurs have solved this problem by having enlarged caeca within which their resident microbes ferment the cellulose at leisure. This is not the best solution because the caecum is rather far down the alimentary canal, which leaves not a great deal of gut wall through which the products can be absorbed. Most folivorous lemuroids have rather good absorptive facilities at the lower ends of their guts but, as explained elsewhere, sportive lemurs solve the problem by passing much of their food through twice.

Cows are well known for chewing the cud and generally taking a long time about processing their food with the aid of a chamber at the front of their stomachs, known as the rumen. Colobine monkeys are not ruminants but their approach to the problem is rather similar. They have chambered stomachs through which large quantities of leafy material are passed very slowly as they are fermented by the gut flora. Thus a colobine has the greater part of its alimentary canal from which to absorb the products of the microbes' work. The microbes also feed and multiply in the process, otherwise they would not be so helpful – but that is another story.

So far, so good: the folivorous primates have evolved ways of unlocking the nutrients in leaves. Unfortunately, the leaves have been forced to take steps to protect themselves from the onslaught of folivorous animals, and although insects present them with their biggest threat (think of a horde of locusts, or a plague of caterpillars, for example), the ways in which plants defend themselves affect any animal that tries to eat them. What follows presumably applies to folivorous lemuroids as well as to colobines, but more evidence has been collected in relation to the latter.

Since colobine monkeys are capable of eating leaves, it might be thought that they would be 100 per cent browsers. There are, after all, plenty of leaves around. It turns out, however, that they all eat at least some seeds and fruit; and several species eat very many more of these than they do leaves. 'Topping up' on fruit makes sense because fruits in general contain plenty of readily digestible sugars, but when the food available to any colobine is analysed chemically, it further turns out that no monkey simply goes through the forest selecting a balanced diet. They are prevented from doing this by the plants' ability to defend themselves.

Trees, like other plants, depend on their leaves for the photosynthesis that provides their nourishment. Without their leaves, they cannot use sunlight to obtain food from the carbon dioxide in the air. It follows that a tree that is stripped of all its leaves will eventually die, and a species to which this happens too often will eventually become extinct. It is hardly possible for trees to hide from leaf-eaters, so most of those that have been successful in the rain forest have done something to make their leaves undesirable as food.

They have achieved this by developing an enormous number of different chemicals that are stored within their leaves and which must be eaten if the leaf is eaten. Some of these chemicals act to deter eating before it takes place. Stinging hairs, as on nettles, for example, are a deterrent to sensitive mouths; and nasty-tasting, sticky latex may be equally effective.

The majority of plant defence chemicals, however, exert a bad effect on a browsing animal after they have been eaten. Some simply ruin its digestion by such tricks as de-activating the enzymes that break down complex food or by killing the gut flora. These include fibres, as they seem to inhibit digestion. Others are more generally poisonous and get into the animal's body and may even kill it.

Both kinds of defence chemicals are exceedingly complicated and there are very many of them, because the trees and the browsers are locked in a never-ending evolutionary race. As the trees produce new defence chemicals, so the browsers evolve ways of dealing with them. At the simplest level, there is evidence that most monkeys tend to avoid leaves and other plant material that are defended in this way, but that colobines can tolerate the kinds of chemicals that spoil their digestion if there is enough nourishment in the item, usually in the form of protein, to make eating it worth while. In other words, there is a pay-off between costs and benefits.

There is no doubt that colobines are well in advance of other monkeys and apes when it comes to their ability to survive eating poisonous leaves, seeds, and so on. Some specific examples will be quoted later on. What is less certain is how they do it, but current theory suggests that the gut microflora are active in rendering poisons harmless.

The colobine subfamily should, therefore, be seen as a group of monkeys that excel in their ability to eat foliage. This ability has been the basis of their strategy for survival and has permitted them to become widespread, but it does not mean that they do not eat nourishing, non-poisonous foods when they get the chance.

There are six colobine genera, of which only one, *Colobus*, is found in Africa. The remaining five are Asian.

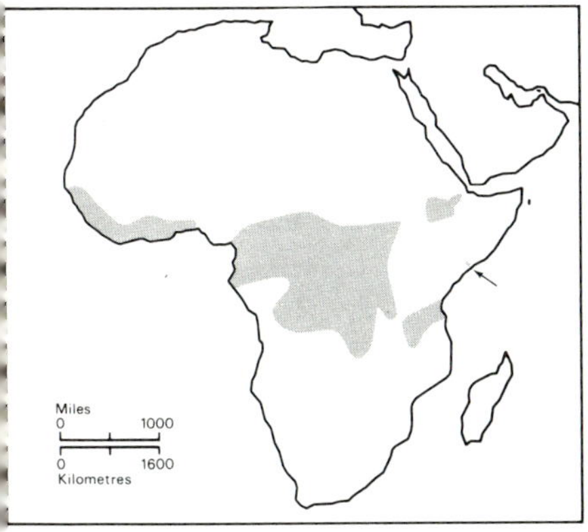

The colobus monkeys GENUS *Colobus*

For a few seconds, a vibrating roar shatters the stillness of the African night. To the uninitiated, it seems that the owner of such a voice must be some large and dangerous beast, but it is only a monkey calling overhead. The male guereza cannot rival the howlers or gibbons for the distance that he can call, but his tenor vibrato carries easily beyond the confines of his small territory.

The 20 or so races of black-and-white colobus monkeys form a distinct group within the genus *Colobus* and are usually put into four species, of which the guerezas are the best known. In their natural habitat, these beautiful creatures seem vaguely improbable

For illustration of guerezas (see page 115).

These three female western red colobus monkeys spend much of their time quietly resting and grooming each other as they digest their rather bulky food.

as they canter along large branches or hurl themselves into the air, crashing with an apparent lack of discrimination into masses of foliage and small branches. Their black bodies are trimmed with long, white mantles and white plumes at the ends of their tails, which billow and swirl as they plummet across gaps in the canopy. Like all monkeys, they miscalculate occasionally, but the vegetation usually breaks their falls and, quickly grabbing hold of small branches, they run up towards safer support as the foliage springs back.

Guerezas start life with a complete coat of short white fur, called a natal coat because they are born with it and it distinguishes them from older animals. At this stage in their lives, the lack of other bodily adornment makes the hooked nose that characterizes the species seem particularly prominent. By the time it is about four months old, the young guereza develops black fur on the top of its head and all over its body and limbs, apart from the long white fringe and tail plume. The various races of guereza differ in the amount and length of the white hair they have, and some aberrant individuals in Kenya go to the extreme of being entirely white. However, this is the result of a single genetic mutation that has spread in a limited number of animals, and they are not regarded as being a separate race. In Ethiopia, brown-and-white individuals have been seen.

Guerezas are found in various types of forest across the middle of Africa, from large tracts of evergreen rain forest, to montane forest patches and strips along rivers in the savannah zone. The males weigh anything from 9 to 14.5 kg (20 to 32 lb) and generally one of them attaches himself to a harem of females, plus their young. The females weigh only between 6.5 and 10 kg (14.5 to 22 lb). Sometimes, there is more than one adult male in a group, but this is unusual.

Their success lies very much in their ability to subsist on a monotonous diet largely made up of the young leaves of the commoner trees in the forest. They play the leaf-eater game to the hilt and can even manage with a lot of mature whole leaves in their diet for weeks at a time, if necessary. It follows from this that their food is all around them; they do not have to travel from one part of the forest to another just to collect a particular item that they need on the menu. In addition, they seldom have to travel to find drinking water because the digestion of large amounts of cellulose releases a lot of water from their food. As a result, groups of about 8 to 15 guerezas can live entirely within about 15 hectares (37 acres) of forest, give or take a little according to the quality of the habitat. Such an area is small enough to be defended and, although they are not always successful in keeping rivals out, guereza groups do attempt to mark out a territory for their own exclusive use.

The males are generally fiercely antagonistic to one another, and sons leave their parental groups as they near sexual maturity. From then on, each must live as a solitary until he can oust another male and take over a harem of his own. Although their lives are fiercely competitive in this way and they can be extremely active at times, guerezas are typical of leafeaters in spending large parts of their days in a somnolent doze during which they can digest their bulky food. They generally manage to travel about 500m (1640ft) between dawn and dusk, but they may not visit parts of their range for months at a time.

Black-and-white colobus monkeys share many of the forests in which they are found with a very different member of their genus, the red colobus. There are about 14 races of these monkeys, but all are usually grouped into a single species. These two types of colobus monkey are able to coexist in the same areas because they exploit the environment in quite different ways. In the Kibale Forest in western Uganda, guerezas and red colobus monkeys eat quite different items for at least 84 per cent of their diet, and what overlap there is concerns common foods for which there is no need to compete.

The red colobus monkeys are rather selective feeders. They move about in home ranges of approximately 35 hectares (86.5 acres), which is more than twice the size of guereza ranges in the same forest. They need this large area to feed in because, although they eat big quantities of young leaves, just like the guerezas, red colobus monkeys cannot fall back on large amounts of common mature leaves when others are in short supply. At best, they can eat the petioles of those leaves, but they must at times search more widely for their food. Thus, even a lone monkey needs to forage in a large home range and a small patch of forest will seldom sustain even a small group of them.

Yet it seems that, because their food grows in sizeable clumps in different parts of the forest, an area that includes the whole diet will be big enough to sustain several quite large groups. This may be why red colobus monkeys form groups of between 19 and 80 individuals, and most of them make no attempt to defend territories from each other. This is not to say that all is peace and love between the groups. On the contrary, when two groups meet, one usually gives way. No doubt the monkeys constantly assess the strengths and weaknesses of their neighbours, but when social conditions are stable, the outcome of any encounter between two specific groups is the same no matter where in the range it occurs. Just to confuse the issue, there are some places in Africa where the local red colobus groups do defend territories. In these cases, they can find all their food in a relatively limited area.

Whereas among guerezas it is the males that leave their groups

LEFT Kirk's red colobus is a very distinct race that is found on the island of Zanzibar.

OPPOSITE Mother and infant red colobus monkeys. This race comes from the far west of Africa, but the species is spread right across the continent.

when they grow up and the females remain, among red colobus monkeys, it seems to be the rule that females transfer between groups, and they may do so several times during the course of their lives. Each group contains several adult males that live together as a cooperative clique. They and the nearly mature males that have grown up in the group handle all the encounters with other groups, which are therefore essentially male disputes. A maturing male must either join that clique or be forced out of the group, as happens to guerezas of the same age. If he is forced out, he will face the problem of finding an opening in another group if he is to avoid a lifetime of solitude.

In at least one race of red colobus, it can be fatal for males to attempt to join new groups. In a rather isolated patch of forest in Gambia, females of the westernmost race have been known to gang up on, and even kill, strange males that made friendly overtures. Gambian red colobus monkeys have a rather spider monkeylike society in which the groups split up into small parties for long periods of time but, as with the spider monkeys, each individual can have little doubt as to the group to which he or she belongs.

Altogether, red colobus monkeys probably have one of the most complex of monkey societies, but it is readily apparent that the life of a female is much more pleasant than that of a male. This is in spite of the fact that males, at 10 to 12 kg (22 to 26.5 lb) are considerably heavier than females in most races. The latter weigh about 8 to 9 kg (17.5 to 20 lb).

The black-and-white colobus group provides another interesting variant on the colobus theme. This is the black colobus monkey, which is in fact all black, although its group affinities are clear. In southern Cameroon, these primates live in a forest that grows on extremely poor sandy soils. This has had an important consequence for the monkeys because there has been tremendous evolutionary pressure on the trees not to lose any more than necessary of their leaves, since it takes nutrients that ultimately come from the soil to produce them. As a result, trees with extremely poisonous leaves have done well and even the black colobus monkey, 'leaf-eater' though it may be, cannot eat large quantities of them. It therefore specializes in seed-eating. Actually, the seeds are probably not much less poisonous than the leaves, but the monkeys can cope with them because seeds tend to be rich in stored carbohydrates, oils and other nutrients that are intended to nourish the young plant. These foods can equally nourish the black colobus monkeys which thereby gain enough to put up with the toxins.

Nevertheless, they are extremely selective feeders, rejecting most of the leaves and seeds in their forest and just concentrating on a minority with which they can deal. One of the tree species they avoid like the plague protects itself not with poisons but with a colony of stinging ants that live in its hollow branches. This is the *Barteria* tree and its ants are fiercely protective of it. Since their stings are more like those of scorpions in their effect than those of honeybees, no monkey in its right mind will tangle with them. However, on the rare occasions that they find a *Barteria* tree without an ant colony, the monkeys will strip it bare of its leaves within hours. This shows just how important it is for a tree to defend itself against the depredations of vegetarian animals.

Apart from the black-and-white and the red groups of colobus monkeys, there is another member of this genus which is given a rather separate status all to itself. This is the little-known olive colobus monkey of the West African forests. It is a much smaller animal than the others: both sexes weigh about 3 to 4 kg (6.5 to 8.8 lb), the females averaging slightly less than the males. It is a short-haired animal with olive-grey fur that makes it fairly inconspicuous in its preferred habitat of dense undergrowth and the dark, lower levels of the forest.

One thing that all colobus monkeys have in common is a thumb that is either absent or reduced to a small protuberance. This is how they get their name: 'colobus' is derived from the Greek word meaning 'docked' or 'mutilated', an allusion to the fact that the thumb looks as though it has been cut off, leaving just a stump or, in some individuals, nothing at all. This in no way handicaps them since they can hold objects between fingers and palm, and a smooth, hooklike hand is the best sort for rapid swinging on branches.

Colobus monkeys have long been successful inhabitants of Africa's forests, but the black-and-white group have been persecuted by man more than have most primates. Like many monkeys they have had to put up with being hunted for their meat and, in recent years, the destruction of their forest home has reduced their geographical distribution; but in addition to this, black-and-white colobus skins have long been in demand for human adornment. East African tribes, such as the Masai and Watutsi, make traditional headdresses and capes from guereza pelts. Much more seriously, there has long been a massive demand in the industrial west for colobus skins for rugs or fashionable garments, revolting as this seems to anybody who has ever watched these delightful creatures when they are alive. The demand was particularly high in the late nineteenth century, when tens of thousands of skins were traded every year, but it still persists to this day; and in some parts of Africa tourists who wish to may contribute to the extermination of these creatures by buying souvenirs that are made from their dead.

The langurs GENUS *Presbytis*

For the langur distribution map (see overleaf).

Travel far on the Indian subcontinent and you will soon become aware of the big grey monkeys with long tails that hang about at the edges of small towns and villages, or even in the bazaars. They have black faces and move with an easy, athletic lope. These are the sacred or Hanuman langurs of the Hindu tradition, often known more prosaically in the West as grey langurs. They are something of an aberrant representative of their genus, being the only one of 16 species that spends up to 80 per cent of its day on the ground and is more at home in open woodland than it is in the moist, green jungles that cover parts of southern and southeastern Asia. But here they are at an advantage over the others, for much of India's forest cover has been destroyed within historical times and yet they have survived, protected by the Hindu proscription against harming monkeys. For this they can thank Hanuman, the monkey god, who aided the god Rama in recovering his beautiful wife from a demon king. Sita was the unfortunate goddess who had been abducted by the awful Ravana and carried off to the ancient kingdom of Sri Lanka. In spite of the fact that her rescue involved such difficulties as a battle between Hanuman's monkey legions and ferocious sea

very arid, open habitats where the bigger groups are most often found. Home ranges also vary greatly between different locations: anything from 20 hectares (about 50 acres) to 1270 (about 3140 acres) is possible. Nevertheless, an average group of Hanuman langurs could be said to range over about 40 hectares of land. Because of the great variation in the type of habitat that langurs occupy, there are no easy correlations to be made between group size, population density and home range. What can be said is that groups will encounter each other most often in the areas of high population density where the single-male harems tend to occur.

The amount of exclusive territory that any one group has to itself is also very variable from area to area, but relationships between groups are generally antagonistic, with chasing and ferocious intergroup fighting being quite common occurrences. This is but one expression of competition between individuals, where alliances are working within the groups.

In the areas of high population density where there are harem groups, the excess males form bachelor bands that roam over very wide areas, and are presumed to be quite unstable in their composition. They lack any truly permanent members, something that the cadres of females provide for the mixed groups. These bachelor bands are best seen as temporary alliances of males that are united by their similar circumstances. They all lack mates. Thus, the male half of this type of langur society consists of an explosive mixture of 'haves' and 'have-nots'.

The harem males are in constant danger of being attacked and displaced by the bachelors; and this is exactly what happens. For every harem male, the time comes when he cannot withstand an attack, either of a single rival or of a whole band of them. In the latter case, a fairly chaotic, albeit brief, period may follow his demise while the members of the bachelor band fight until one of them is left in undisputed possession of the harem. This, then, is the basis of the harem cycle: possession and usurpation every two or three years, while the females remain in the group. There are, however, considerable variations on the theme, and lone males quite often manage to hang about with others' harems on a temporary basis. They may spend their time at the periphery and occasionally even sneak in and copulate with one of the females. More often, lone males or bachelor bands form a very obvious presence close to the harem and harass the resident male in a war of nerves.

Whereas in border encounters between two whole groups, the females and even some of the young animals will join the dispute, it is left to the harem male to drive off bachelors that try to sneak in. The adult females do, however, join in the fray when invading males make serious attempts to usurp their harem lord. They then may attack and even kill the invader.

In so doing, they appear to be protecting themselves because the males have escalated their own competition to the point where it could severely damage the interests of the females. Given the opportunity, an invading male will follow up a successful take-over by killing all of the unweaned infants in the group. These infants are, after all, the offspring of his predecessor and therefore competitors with his own as yet unborn issue. More importantly, their continued existence would prevent his access to their mothers because the females do not become sexually receptive during the 12 to 20 months that they spend nursing. But when a female loses an infant, it is in her best interests to become pregnant again as quickly as possible.

It is therefore adaptive for a male to kill off his predecessor's offspring because in doing so he not only eliminates some of the competition but, more importantly, he speeds up the availability of receptive females and therefore the advent of the next generation in his line. Since his successor will attempt to do exactly the same thing in his turn, speed of reproduction is essential if any of the young are to be safely weaned before the next coup.

The cycle of take-overs and infanticides is not as simple as it might be because, in spite of their relatively small stature, the females are often able to break it by cooperatively defending their babies. A male cannot simply kill infants with impunity. He must snatch them in a moment of maternal slackness or vulnerability in order to deliver the fatal bite; and natural selection has favoured those females best able to protect their young. A mother may even go to the extreme of leaving the group in order to take her baby away from the danger.

It might be thought that a female would refuse to mate with her offspring's killer but if she is to remain with her group, she has little option. And she does at least gain a proven, competitive male as the father of her next offspring. She has nothing to gain by choosing to mate with an effete, non-infanticidal, non-competitive individual since this would only increase the chances of her having non-competitive male descendants. In the numbers game of evolution, this would make her a loser.

And so it is, that within a few hours or days of an infant's death, its bereaved mother will be presenting her rump to her new mate and signalling her sexual willingness with the species' characteristic shuddering movements of her head. A female langur that is neither pregnant nor lactating normally has a 28-day menstrual cycle and is probably most receptive at about mid-cycle, when she ovulates. However, the upheaval of take-over and infanticide accelerates her sexual enthusiasm. Whether or not it also accelerates ovulation is not known; it may be that it is in her interests to consummate the relationship with her dangerous new lord as quickly as possible, in

Adult female Hanuman langurs combine to chase a big male that threatens their infants. Infanticide by males is common among langurs.

order to calm him down. If she is already in the early stages of pregnancy, whether or not she has lost an infant to him, it may be in her interests to establish herself as his mate, should he have a general inclination towards killing non-mates' offspring. This, however, is getting rather far into the realm of speculation, with all sorts of implications about males' abilities to respond to the time interval between mating and birth, and appropriate data do not exist.

As a postscript to the story, no one should be surprised to learn that male Hanuman langurs seldom reach adulthood in the troop of their birth, often making a hasty exit as juveniles at the abrupt end of their father's tenure. Male take-overs and infanticides also occur among other monkey species, including the silvered langurs

This banded langur comes from the centre of the Malay Peninsula. The species if widespread in South East Asia, and other races may be different shades of grey, or even reddish in colour.

RIGHT **This baby silvered langur will grow up into an adult with fur that is a mixture of light and dark shades of grey. The orange 'natal coat' only lasts for a few months (see page 132).**

BELOW **The pale rings around its eyes give the spectacled langur its name (see page 132).**

of South East Asia and Sri Lanka's purple-faced langurs. Among the latter, juvenile females as well as males sometimes leave during take-overs and spend a while with the roving bachelors before settling down to more stable group life.

Apart from the Hanumans, the other 15 langur species are highly arboreal monkeys which tend to have large amounts of foliage in their diets. Much as with Africa's colobus monkeys, there are some ecological differences between the species which often allow two of them to live side by side. In certain parts of Borneo, for example, white-fronted and maroon langurs share the virgin jungles; and in the Malay Peninsula, banded and spectacled langurs manage to coexist by means of slightly different feeding and ranging strategies.

The forest langur species tend to be smaller than the Hanuman, with less of a size difference between the sexes. For example, male spectacled langurs weigh about 7.4 kg (16.3 lb) to the females' 6.5 kg (14.3 lb). The forest species also tend to occupy smaller patches of forest, 30 hectares (75 acres) being about the size of a home range for a banded or spectacled langur group. There is, of course, a considerable amount of variation and some groups of Sabah's beautiful maroon langur occupy ranges of 80 hectares (200 acres).

The 16 species fall naturally into four groups, which have at times been accorded the status of subgenera or even genera. If the adults seem to be a rather confusing array of different types, some of their group affinities can be conveniently identified by looking at the distinctive colours of the infants. The banded langurs, maroons and four other species have pale babies with a big, dark cross on their backs; and the silvered, spectacled and five other species in their group have babies covered in startling shades of bright orange fur. The Hanumans and the last group (Nilgiri and purple-faced langurs) have black or grey infants. These striking colours seem to play a role in making the infants attractive to other members of the group; and langur mothers, like most colobine monkeys and unlike several of the cercopithecine subfamily, usually have a very relaxed approach towards allowing other individuals to babysit. Whether or not this is actually good for the babies is open to question and would bear more investigation, as the 'aunts' are often either casual or incompetent, but the genus as a whole is a very successful one.

The proboscis monkey GENUS *Nasalis*

The proboscis monkey is the only member of its genus. It is a large and remarkable animal that is unlikely to be confused with any other monkey. The long, somewhat pendulous nose of the big males is a unique adornment of doubtful utility. It can be neither inflated

nor erected, although it does move about when the mouth is opened or closed, and this is particularly noticeable when the male is calling.

Some scientists have postulated that proboscis monkeys are descended from smaller, snub-nosed monkeys that grew bigger as they evolved over hundreds of thousands of years. It is argued that the nose was biologically programmed to grow faster than the rest of the body, so as proboscis monkeys evolved into larger animals, their noses evolved into huge organs. Since the males are larger than the females, their noses are relatively bigger.

Unfortunately, while this theory allows us to avoid having to explain what a proboscis monkey does with its mighty protuberance, it begs the obvious questions about why a snub nose was there in the first place and why it should grow faster than the rest of the body. It would be more straightforward to explain the big nose in terms of its usefulness, but it looks as if it is capable of doing little other than get in the way. However, this gives the alert biologist an immediate, if somewhat paradoxical clue: if the nose is a nuisance at times, it must be worth having at other times or natural selection would have favoured smaller-nosed individuals over the generations. Perhaps the answer is that proboscis monkeys, especially the females, prefer mates with big noses, so the better-endowed individuals have relatively more offspring and the adornment is maintained, no matter how inconvenient. If this is true, it is a good example of sexual selection.

Newborn proboscis monkeys have deep-blue faces with light rings around their eyes and little upward-pointing noses, but by the time they are about nine months old, their faces have assumed the adult, pinkish-brown coloration. Although the species has only one race, the pelage colour varies somewhat through different shades of brown. It is often almost red on the crown, neck and shoulders, and is generally darker on the back, giving way to a lighter, more sandy colour on the sides and front of the body. The tail and a distinct patch of fur on the lower back are always pale cream or grey.

The males are about twice as big as the females, sometimes weighing very nearly 24 kg (52 lb). The female record is just under 12 kg (26 lb). They are thus among the most dimorphic of monkeys. As a result of the size difference, full-grown males move rather more carefully in the trees than do females and younger males, but all may swing under branches or leap between supports, often landing with a crash in a mass of foliage. They are also given to plunging from the trees into open water, sometimes dropping from as high as 16 m (53 ft). They are excellent swimmers, able to dive and swim under water to evade danger.

Proboscis monkeys are found mainly in coastal swamp forests and alongside the rivers that flow through the flat, coastal plains of

Borneo. There are occasional reports of individuals much further inland at higher elevations, and groups will readily climb hills near the sea. However, it seems that the groups always start and finish the day in trees that jut out over water. It may be that these offer protection against their main natural enemy, the clouded leopard.

Proboscis monkeys have neither been adequately surveyed as to their distribution in Borneo, nor have their ecology and behaviour been studied in any great detail. Their society appears to be based on groups of 10 to 60 or so animals, each containing a few large males, rather more adult females and a variety of juveniles and infants. The groups do not appear to defend territories and may be socially more flexible than most primate societies in that individuals may transfer easily between them, or the groups themselves may be only temporary subdivisions of a larger society, subdivisions that may coalesce and fission from time to time.

The group's day begins when the monkeys wake up at dawn in a large tree over the river or swamp. As individuals become more active, so they move out in ones and twos to the nearest source of palatable leaves, young leaves being preferred. They then pass the day eating foliage, digesting it and moving on to the next food source. Like all leaf monkeys, they can spend hours apparently doing nothing after a good feed.

Their postures in the trees are sometimes remarkably human, as they sit propped in a convenient fork, dozing or bending accessible small branches towards themselves and stripping them of their leaves. Towards the end of the day, the progression of the group tends to speed up, especially if by then it has moved rather far from the river or swamp; and as dusk falls, the monkeys settle down to sleep in the same or another tree over the water, having travelled a kilometre or two since the break of day.

This quiet life is occasionally punctuated by squabbling within the group or by an external threat, such as the approach of a human. In both cases, the males then play an important role, using loud snorts and honks to admonish squabblers or to threaten an approaching person while the rest of the group flees.

Proboscis monkeys may occasionally be preyed upon by clouded leopards; and crocodiles, pythons and perhaps even black eagles may take individuals once in a while, particularly the infants and juveniles. Such predation is no threat to the survival of the species. Subsistence hunting by the people of Borneo is much more serious. In addition, the proboscis monkey has the misfortune to be a riverine and coastal species in a part of the world where human development is rapidly spreading up the rivers and where coastal forests are often demolished on a large scale for timber products. Thus its long-term future is likely to be entirely dependent on the protection offered by a few suitable parks and reserves.

LEFT There is no parallel elsewhere in the monkey world with the enormous nose of Borneo's adult male proboscis monkey, although the local people like to refer to it as 'orang belanda' or 'Dutchman'.

BELOW The simakobu is one of the least known of primates. This adult female has golden hair, but many individuals have dark grey pelage.

The simakobu GENUS *Simias*

For the last half million years or so, a deep ocean channel has isolated the Mentawai Islands of Indonesia from the mainstream of primate evolution. While frequent changes in the sea levels around Southeast Asia have, from time to time, permitted primates and other animals to migrate within much of the region, those on the four Mentawai Islands, just to the west of Sumatra, have followed their own unique evolutionary pathway. The gibbons, the langurs and the macaques are each represented by a single species that is not found elsewhere, and the simakobu is so individualistic that it is given its own genus, peculiar to the Mentawais alone.

It is unique among the colobine monkeys in having a short, pig-like tail and two distinct colour varieties: some individuals are dark grey, others are a beautiful, pale golden colour. Even within a single family, the two colours occur and 'family' is an appropriate term to apply to the simakobu's domestic arrangements. Although larger groups have been seen, the usual convention is for pairs of adults to live together with their offspring in territories of up to 50 hectares (124 acres) that they seem to defend against others of their species. When two families meet, at their common boundary, the males call loudly but briefly to each other in what may be a rather simple ritual, then each dashes back to the centre of his own range, closely followed by the rest of his family. The brevity of these encounters is in keeping with the generally rather low-key tenor of the simakobu's existence. It spends most of its time either sleeping or foraging for leaves (and some fruit) high in the trees of the tall, mature forest.

When alarmed, it makes no noise, but either hides in the foliage or flees as rapidly as possible. The local people favour simakobu for their meat and have been hunting them with poison-tipped arrows for hundreds, if not thousands, of years. In response to this, it has behoved the monkeys to become expert at concealment, so that spending a couple of hours in frozen immobility surrounded by leaves is something that they do very readily. Surprisingly for such arboreal monkeys, simakobus will also shin rapidly down tree trunks or leap from great heights to the forest floor, where they slip away through the densest undergrowth they can find. No doubt the golden individuals are easier for the hunters to see than their dark fellows, but in some places they have special protection as a result of a local belief that they are sinister, rather ghostly beings that it is unwise to hunt.

Simakobus rarely call during the normal course of the day, lacking any habit of regularly giving out loud territorial advertisements as do so many other species. Nevertheless, the males do possess the capacity to give a bark that is audible several hundred metres away, to which the females and young respond with shrill squeals.

Borneo's proboscis monkey is the simakobu's nearest relative. Although the adults of the two species do not seem to be much alike, the simakobu has a tipped-up, snub nose that is almost exactly the same as that of a newborn proboscis monkey. The former animal is, however, very much smaller than its big-nosed cousin, being shorter in its limbs and probably only about half the weight. The single individual whose weight is on record is a female of 7.2 kg (15.7 lb).

The two genera may have a common ancestor that lived not long before they became isolated on their respective islands, since when the very different conditions in which they live have brought about an extremely rapid divergence in anatomies and life styles.

The snub-nosed monkeys GENUS *Rhinopithecus*

In the high mountains of western China, amid the forests of pines and rhododendrons, the winters are bitterly cold and even the summers are not very hot by most monkeys' standards. Snow covers the landscape for more than half the year. Nevertheless, this is the home of the golden snub-nosed monkey, a stocky primate that has a luxuriant coat to combat the elements. And perhaps because of the cold, males and females spend hours hugging each other, with their faces buried in each other's fur and their long tails wrapped around the outside of the warm bundle. When the female needs a hug from her partner, she walks up to him and gives him a friendly butt with the top of her head. It does not seem to be in his nature to refuse very often, but then life would be tough for a real loner in the cold.

Like many mammals that live in highly seasonal climates, golden snub-nosed monkeys change their coats annually; they moult as autumn approaches and a new coat, redder than the old, grows in before the winter. They also breed seasonally, as far as is known, mating in the early part of the winter and giving birth in the spring. The babies have very pale fur for the first two years of their lives, but they darken to the adults' colour by the age of two and are thought to reach full maturity by the time they are four.

Both sexes have big, soft, blue muzzles, with bare, bluish skin around the eyes and up-turned noses that are like those of the simakobu. The tips of their noses point up so far that they almost reach their foreheads. The adult males have extra adornments in the shape of fleshy flaps on either side of the upper lip, the function of which is unknown. The latter are also bigger than the females, weighing about 15 kg (33 lb) to their 9 or 10 kg (20 to 22 lb).

The ecology and social behaviour of these spectacular animals are largely unknown, but their robust bodies and relatively short limbs suggest that they spend a lot of their time on the ground. They are said to descend in the winter from the high mountain slopes into the warmer valleys, so the ability to cross open, treeless areas might be important when they are travelling.

Since they are colobines, they probably include many leaves and shoots in their diet but the very harshness of their environment suggests that they must be prepared to capitalize on whatever is available, and it seems reasonable to expect them to eat bark, pine needles and cones (which over-winter in a green state) during the cold months.

There are three races of golden snub-nosed monkeys in western China and one more member of the genus, the Tonkin snub-nosed monkey, in northern Vietnam. The latter is a smaller, darker species living (if it still exists) in tropical forest.

For illustration of golden snub-nosed monkeys (see page 138).

Golden snub-nosed monkeys inhabit the bitterly cold mountains of western China. They spend much time hugging each other, perhaps for warmth, and carefully grooming each other's fur. Here, a female grooms a male.

The douc GENUS *Pygathrix*

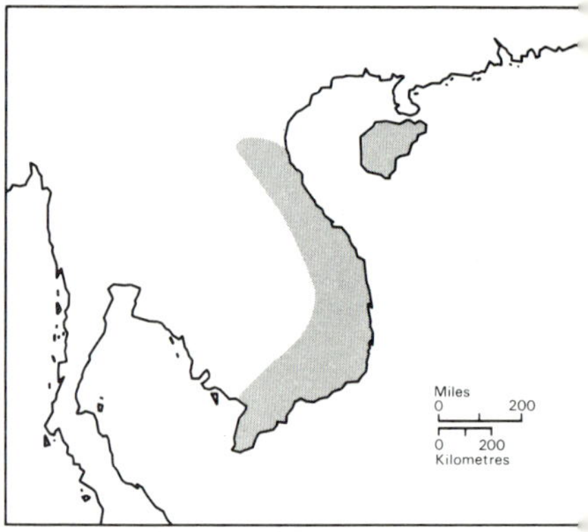

If ever there was a monkey that deserved a break, it is the douc (pronounced 'dook') of South East Asia. During the height of the Vietnam war, literally millions of bomb craters were being made every year in its forest home and, as if that was not enough, huge tracts of land were defoliated with chemicals, effectively wiping out the natural communities of plants and animals. On top of the destruction of their natural habitat, doucs have had to put up with being used for target practice by trigger-happy soldiers and for meat by subsistence hunters. Nobody knows how many doucs survived these horrors, but it seems likely that their numbers have been devastated over large areas.

Doucs are among the most colourful and attractive of all primates. They are about the size of the biggest langurs, perhaps a little bigger, and the females are only slightly smaller than the males. There are two races, the more northerly of which is the

Doucs are among the most brightly patterned of monkeys but they are quiet in their behaviour. They depend heavily on leaves for their diet and most spend long periods of time gently digesting their food.

better-known one. It is referred to simply as the 'douc' and has a naked, yellowish face that turns darker if the animal spends much time in the sun. The almond-shaped eyes have a distinctly 'Asiatic' appearance but it is sheer coincidence that an Asian monkey should resemble the people of its region in this way! The eyelids are a soft powder blue. A fine beard of long white hairs rings the face, and this is considerably bigger in the males than in the females. The body hair has a greyish 'pepper-and-salt' appearance; the thighs are black, and the top of the head and legs below the knees are deep maroon. The tail is white and has a distinct triangle of white hair at its base, at the top corners of which the males have two conspicuous white tufts of hair. Broad white cuffs cover the wrists, sometimes up to the elbows. The overall effect reminds many Westerners of a particularly outlandish pair of pyjamas.

The black-footed douc of southern Vietnam is similar except that it has a dark face, all-black legs and lacks the white cuffs.

Although very little is known about the daily existence of wild doucs, they appear to live in groups of less than a dozen animals, in which the females outnumber the males. They inhabit both undisturbed, mature monsoon rain forest and areas of younger growth that result from human disturbance of the habitat. When a group is relaxed, it progresses through the forest in an extremely noisy fashion, crashing through the foliage, swinging under branches, leaping bipedally with a remarkable show of balancing ability and sometimes covering gaps of 5 or 6 m (nearly 20 ft) by means of diving leaps. In the last case, the landing is made feet-first, like a sifaka. On the other hand, when a group is disturbed, it can flee silently through the tree tops.

Like most colobines, doucs eat large quantities of leaves which they supplement with fruit. They are fairly chaotic feeders, dropping huge quantities of food to the forest floor, but they are peaceful with each other and do not seem to squabble over their food. Often, they share the same clump of foliage and may even break pieces off and hand them to each other, a type of active generosity that is rare among Old World monkeys.

In spite of the noise they make when travelling, most of their life is rather quiet and long periods are spent silently digesting their bulky food. When startled, they can give loud barks and dash around in the trees slapping the branches with their hands and feet; but dozing and quietly grooming each other's fur take up much more of their time.

When they play, they adopt a distinctive expression with the mouth open, teeth partially bared and chin thrust forward. Sometimes, they close their eyes and paw blindly towards each other with remarkable disregard for the hazards of doing this when up a tree.

In recent years, these monkeys have been referred to as 'douc langurs', but in fact they are much more closely related to the proboscis monkey and the snub-nosed monkeys than to the langurs. The word 'douc' is an ancient name of Vietnamese origin.

II The Generalist Monkeys of Africa and Asia: intelligent manipulators

These are the cercopithecines, the hustlers of the primate order. They have managed to colonize some of the most unlikely primate habitat, including arid scrubland, open montane grassland and temperate forests that are snow-covered every winter. Between them, they can eat just about anything that is edible to any primate, with the exception of mature leaves.

A few species have become rather specialized, but most of the rest are notable for their adaptability. In a single day, a monkey might consume whatever it can obtain in the way of fruit, flowers, young leaves, insects and even meat; and it might range from the tops of the tallest trees right down to the ground. It is this adaptability that has allowed the Old World generalists to be so successful and so widespread, because only the most inhospitable environments have checked their advance, and they have been able to survive in very seasonal habitats where there are periodic, sharp changes in the weather and in the type of food that is available.

They have simple digestive tracts that can cope with a diet as varied as that of a human, but they have added a unique refinement: large extensions of the cheeks form pouches that can be packed with food, such as fruit, which is often collected quite some time before it is eaten. This allows them to load up with food in what may be a dangerous place, before retiring to some more secure location for the actual business of eating. Dominant individuals sometimes make an extra use of the system by forcibly emptying the cheek pouches of weaker monkeys, once they have gulped down their own share of the food.

Many cercopithecine species live in social groups that have a rather authoritarian internal organization, with bluff, violence and convenient alliances being brought into play to settle disputes. The development of their intelligence has been particularly important to species in this group, as much for individuals to cooperate with and to outwit each other as for them to cope with learning about their complex environment. All in all, they tend to be more active, quick-witted species than are the leaf-eaters.

The guenons GENUS *Cercopithecus*

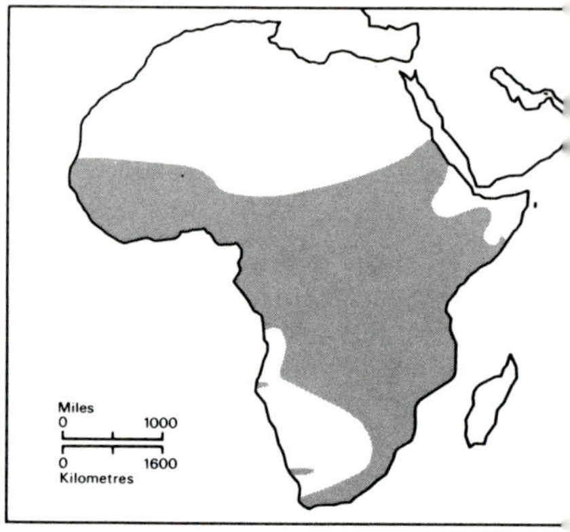

Some 25 monkeys sit in the shade of a clump of fever trees in Kenya's Amboseli National Park. They are rather dull-coloured creatures with greyish-brown pelages and black faces fringed above with a white stripe and at the sides with white whiskers. The adults among them weigh about 5 to 7 kg (11 to 15.5 lb), with the males being quite noticeably bigger than the females.

One male gets up and stalks confidently through the grass with his tail held vertically upwards. He approaches a seated male who hunches down, head held low. The first monkey now parades backwards and forwards in front of the second, all the while orienting his rear-end towards the crouching one. Every now and then, there is a pause in the parade and the monkey stands tail up, bottom towards the seated one, displaying his red skin by the base of his tail, a patch of white fur below that, and then his striking, bright, powder-blue scrotum. He is giving the stereotyped red-white-and-blue display, asserting his social dominance over the crouching individual, who is by now shuffling backwards and giving catlike wails through bared teeth.

For illustration of a female vervet and her infant (see frontispiece).

These are vervet monkeys, East African representatives of the green monkeys, a diverse collection of primates that inhabit savannahs and woodlands all over the continent and make up what is probably the world's commonest species of monkey. The red-white-and-blue display is a sophistication of some of the vervets, by which dominant males remind subordinates of the *status quo*. They do it most often in the mating season, probably to reinforce their personal

Vervet monkeys frequently stand upright during social encounters. The brightly coloured genitals of the male make a conspicuous signal.

rights to the females in the group, since a major advantage of male social superiority in vervet society is more frequent sex with the females. The importance of this, of course, is that a dominant male is likely to have more offspring.

Throughout much of Africa, in the country that lies between the desert fringe and the forest zone, the different races of green monkey – the only savannah-dwelling guenon – have a similar social organization. Each group contains several adults of either sex (usually at least twice as many females as males), plus their offspring. Each male competes to cow the others into submission by means of threats or fighting, so that in the end a dominance hierarchy is formed. The fighting that goes on, particularly in the mating season, can be fierce and wounded male monkeys are a common sight. Deep flesh wounds often occur and injuries to genitals are not uncommon. In the latter case, the winning animal may have eliminated a sexual rival for all time, an unusually direct expression of competition among primates. Whether or not such fighting is good for the species, or even the social group, is neither here nor there; natural selection rewards the male who fathers most offspring, and if attacking rivals is the way to do it, that is what he will do.

As is so often the case, female society among the green monkeys is rather more pleasant with a general air of collaboration replacing that of competition. This might be due to the fact that females stay in the one group all of their lives and are therefore usually quite closely related to each other, whereas adult males are always immigrants from other groups. In theory, closely related animals should help each other more because they share much of their genetic make-up and are thus helping their own genetic lineage.

The females all tend to become sexually receptive to the males' advances at the same time of year, with the result that births are grouped highly seasonally. This is something that is very much under the control of the female of the species since males attempt to copulate at any time of year, but are usually rewarded with a quick slap, or at least a refusal, outside the mating season. Clearly this does not satisfy the males, who are inclined to masturbate when the females are not in the season to be compliant.

By no means all monkeys are seasonal breeders and it may be that the green monkeys have become so because the savannah is a much more seasonal habitat in terms of food production than the rain forest. However, this cannot be assumed and an alternative explanation is that it is the synchrony itself that is important. Babies are a real focus of attention within the group, with other monkeys almost constantly coming around and attempting to hold the little one, or grooming the mother, ever closer to the clinging bundle of dark fur, until that is what is being groomed. This goes on no matter how often the mother moves away or turns her back on her offspring's admirers. Although it may be useful to have a baby-sitter once in a while (as among the squirrel monkeys), some of the attentive 'aunts' are inept young females that can only distress or even endanger the infant. And if a mother does give up her baby, it could take her up to an hour or more to persuade the minder to give it back.

All this must be very stressful to mother and baby alike, so it is possible that each new mother is better off if there are other foci of attention within the group. If that is so, then birth synchrony is an advantage to all, no matter when it occurs during the year; although it would, of course, be best to time the birth season for optimum ecological conditions. The latter could mean a number of things, but might include a rich food supply during the months of pregnancy (for the mothers) or at weaning (for the infants). The only problem with this idea is that different populations of green monkeys have birth seasons that bear quite different relationships to such environmental conditions as wet or dry seasons and times of least or most food availability.

Mother green monkeys have no worries about balancing their babies' use of each breast; their nipples are so close together that the infants usually suck on both at once. Weaning is a very gradual process brought about by the increasing tendency of a mother to refuse her baby. She may pull away or even reward too much persistence with a slap. There is much individual variation in the timing of this, and whereas some mothers start weaning after as little as a month, others even accommodate a year-old infant after a younger sibling has already arrived.

Once past the weaning stage, green monkeys have an eclectic diet. They have a tendency to eat into whatever is available, depending upon the location and the season. Both fruit and invertebrates, such as insects and spiders, seem to be sought after, but certain kinds of flowers and carefully selected young leaves are also taken in considerable quantities. They obviously enjoy eggs and meat, sometimes dismembering birds' nests or chewing on chameleons as if they were toffee apples. When a plague of rats ate up most of the normal food of one West African population of green monkeys, the monkeys responded by eating the rats.

Adaptability is the secret of their success. They cannot compete with true forest species in the forest, but have been quick to exploit that habitat where man has altered it to suit them. The Bakossi tribe of southern Cameroon are a people who practise a mixed form of agriculture on the rich, black soil of their land within the forest zone. Over the last 70 or so years, they have effectively driven a wedge of farmland from the southern fringe of the savannah, deep into what was formerly the habitat of rain forest monkeys. The forest species have retreated, being unable to survive in a constantly shifting mosaic of coffee, cocoa, yams, plantains, oil palms and fallow land. Dense, new forest rapidly regenerates on the latter, but it is very different in composition from the natural vegetation with its tall trees and lack of undergrowth.

Green monkeys do wonderfully well here by using the fallow land for refuge and moving out from this to raid crops. In their natural habitat, they use clumps of tall trees or the forests that line rivers and streams for refuge and move out into the open woodland to forage. The parallel lies in their being experts at crossing dangerous open areas on the ground. Whole groups of them seem to be able to hide behind a few blades of grass; something that more arboreal forest species – even of guenons – just cannot do.

Nevertheless, the green monkeys have had to make some changes in their habits, which is where their sheer adaptability comes in. In the savannah, they have regular ranging patterns, tending to exploit the same food sources at certain times of day in any particular season. In the farmland, they have responded to the Bakossian farmers' attempts to defend their crops by adopting totally unpredictable ranging patterns. They have also become quieter, not by uttering fewer calls, but by making more use of the quieter ones in their repertoire. In effect, they have taken up whispering. That the groups also divide up this new, food-rich habitat into small, rigidly defended territories is not surprising. They usually defend territories in their natural habitat – although being what they are, they make exceptions wherever the distribution of food indicates that they are better off without doing so.

One last adaptation to the new environment is a truly remarkable example of the species' useful flexibility. Whenever the savannah-dwelling ones encounter any hyena or member of the dog family (including people's pets), they climb into the trees and utter loud barks and other warning calls. In the farmland, on the other hand, they hide quietly. In the savannah, of course, doglike animals are terrestrial predators and a monkey in a tree is safe to warn others of the danger. In the farmland, however, dogs are domestic animals that tend to be accompanied by men who do not like monkeys – except possibly to eat. A noisy monkey in a tree would be a sitting duck for a gunshot or a well-aimed stone, so silence is the best policy.

Although it is easy to see that green monkeys are highly adaptable creatures that have made a brilliant job of exploiting a new habitat, it is not so easy to see whether they do it by learning or by instinct. It is just possible that the Bakossian farmers are so effective at weeding out those individuals that behave inappropriately that selection has produced a population of green monkeys with a different genetic composition from those in the savannah. Just as farmers can select strains of cattle, for example, the genes of which dictate that they give more milk, so maybe they can inadvertently select strains of monkeys with genes that make them better crop raiders. In all likelihood, the real answer lies in a mixture of learning and inheritance.

Although green monkeys are so widespread and common, the vast majority of guenon species are forest animals. Some, like the mona and redtail monkeys, have come to terms with agricultural encroachment and can survive in a mixture of forest and cultivation, but most of the 18 species and 51 races are restricted to relatively undisturbed natural forests, ranging from hot, lowland areas to cool, montane bamboo.

The forest guenons are graceful monkeys with long tails. Their silky coats have an exceptionally clean look about them and are often flecked, pepper-and-salt style, in various shades of browns, greys, reds and greens. This flecked effect is achieved by bars of different colours on each individual hair. Their heads and rear ends tend to have conspicuous patterns and colours, which probably facilitate species recognition. Moustaches, pale nose spots and striped sideburns are common, and blue scrotums crop up again.

In many parts of the main African forest blocks, several guenon species live side by side, coexisting by means of fine ecological separations, such as habitually foraging at different levels of the canopy. In western Uganda's Kibale Forest, for example, L'Hoest's guenon forages mainly on the ground and in the undergrowth, while blue monkeys and redtails make extensive use of the levels higher up. It seems unlikely that there is much competition for food between L'Hoest's guenon and the other two, and the latter, in turn, have slightly different specializations from each other. Both are fairly omnivorous, tending to eat a lot of fruit and insects, but redtails specialize in the sorts of insects that need to be caught with a quick grab, whereas blue monkeys go in for the slower, crawling type. Blues also eat more leaves than redtails, although not by any means on the same scale as colobine monkeys.

Not much is known about the social life of L'Hoest's guenon for the simple reason that ground-dwelling monkeys are incredibly hard to spot and to follow in the forest. Arboreal ones are much easier and redtails, in particular, tend to give away their location by scolding humans with high-pitched calls. The two species are easily distinguished by sight: the blue-grey coat of the blues gives them a generally dark appearance in the forest, whereas redtails are more brownish on the upper body, with pale undersides and tails that appear as a startling, almost orangish-red when they catch the light. Facially, too, they are quite different: the dark hair on the blues' heads diffuses into pale grey cheek whiskers that give a 'puffy' shape. Redtails have white nose spots on dark faces and a sharply

ABOVE The white nose-spot and the red tail make it easy to identify a redtail monkey in the forest.

RIGHT The monkey in the middle is a hybrid between a redtail and a blue monkey. In comparison with the redtails on either side of her, she is less brightly marked, and the hair on her cheeks gives her face a more rounded appearance.

defined black stripe in their side whiskers. With males weighing 6 kg (13 lb) and females 3 to 4 kg (6.5 to 9 lb), blues look bigger than redtails. Males of the latter species weigh about 4 kg (9 lb), females about 3 kg (6.5 lb). These clear, physical distinctions between the two types have allowed the detection of an unusual phenomenon, as will be explained.

Both species live principally in harem groups in which a single adult male accompanies several females and young. Redtails move in slightly larger groups but in smaller home ranges: 30 to 35 animals in 24 hectares (59 acres) being about average, compared to about 24 animals in 61 hectares (151 acres). The home ranges are defended as group territories from which other groups of the same species are excluded. Although so closely related, blues do not exclude redtails from their territories, nor vice versa. Territorial defence is not left to the males: virtually the whole group will join in when neighbours test their boundaries, with the females tending, if anything, to be the fiercest of the guardians.

One of the consequences of a harem society is that there is always a surplus of males, in these cases leading solitary lives but ever ready to challenge the harem leaders for their positions. The harem males tend to be toppled about every two years, although there is much variation in this. This is where the unusual phenomenon comes in, because blue monkey males that cannot make it in their own society sometimes join redtail groups. They slowly become established members, showing a mutual tolerance with the resident redtail male. This is quite different from the frequent, but temporary, mixing of groups of different monkey species that occurs widely in the African forests. The female redtails are slow to accept the alien male, perhaps taking years before they will consent to mate; but when they do, hybrids (crosses) result. These are unmistakable as mixtures of the two species, having some red on their tails (but less than their mothers have) and faces that look like washed-out redtails' faces with the puffy, blue-monkey shape. Rather faint versions of the redtails' nose spots and whisker stripes are discernible. The hybrids tend to be the size of blue monkeys, which makes them look pretty hefty in a redtail group.

Female hybrids do well in redtail society, growing up and living in the one group. They groom with other group members, exchange baby-handling with other mothers as is the norm among redtails, and participate in territorial disputes with other redtail groups. Male hybrids must have a tougher time, because they are not known individually to any of the other groups in the area and may be just as slow to find female acceptance as were their fathers. Of course, we may speculate that, being bigger, they would find it easier to win a harem, but it must be galling if the females then take months to accept the situation.

Blue–redtail hybrids are relatively rare compared with the two parent species, but the success of the females spoils the neatness of the definition of a species. This is because they are definitely fertile and have offspring either by their redtail harem leaders or by blue males that move in just as their fathers did. The line has not been followed beyond that, but it does look as if members of two obviously good species can have common descendants. For the present, it is necessary to accept the fact that the best available definition of a species has its limitations.

Other than in their hybridization, however, blues and redtails may be accepted as being typical of the forest guenons – and it is even possible that other pairs of species cross-breed once in a while.

One rather aberrant guenon that deserves a special mention is Allen's swamp monkey. This is a somewhat stockily built animal with a relatively short tail, that inhabits the soggy forests of a rather restricted area near the lower Congo (Zaire) river – and there is nothing like living in a foul swamp for keeping inquisitive scientists at bay! There are suspicions that when more is known about this monkey, it may require its own genus; but, until then, it is easier to think of it as a thoroughly mysterious guenon.

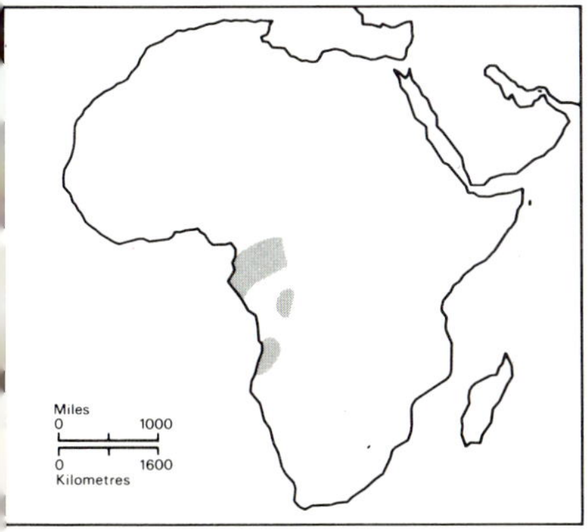

The talapoin GENUS *Miopithecus*

Manioc is a root crop that is quite a common food in the tropics. It is, however, poisonous when first dug up and must be soaked for a period to leach out the toxins and make it edible. In the West African state of Gabon, some villagers do this by soaking it at regular sites beside the rivers, knowing from experience when the manioc is ready. Unfortunately for them, the Old World's smallest monkey, the talapoin, also lives beside these rivers and knows only too well when to nip in first and grab the food. As in the case of the crop-raiding green monkeys that have changed their habits to suit a new life style, the clever timing of the talapoins is all the more remarkable considering that manioc is an introduced plant in Gabon, and although the date of its introduction is obscure, it cannot have been that long ago.

That they plunge into the water to steal food is not surprising because they are excellent swimmers and sleep on thin branches or vines over water every night. If threatened by some sort of cat or snake, they simply drop off their perches and swim away to safety under water. They even have a sporting chance of escaping from aquatic cobras, since these snakes must climb into the tree before the monkeys let go; and it seems rather unlikely that the snakes have worked out the advantage of hunting in pairs, with one up the tree and the other in the water.

Talapoins are not at all closely related to South America's squirrel monkeys but they have evolved to have a rather similar appearance and lifestyle. This is an adult female with her offspring.

The chief reason why talapoins are vulnerable is that they are so small. Males weigh between 1.2 and 1.4 kg (2.6 to 3 lb), females between 745 g and 1.12 kg (1.6 to 2.5 lb). They are not much bigger than the squirrel monkeys of South America and, like them, they live in large groups with the mature males generally rather separated from the adult females and young. Groups may contain over 100 individuals where crop raiding provides a good source of food, but 60 or 70 is more usual in undisturbed, natural habitat.

Among talapoins, the group is clearly the overall, permanent social unit, but it tends to break up into subgroups along the lines indicated, or sometimes into one of juveniles accompanied by a single adult male. In the latter case, the male seems to be something of a baby-sitter who prevents the little ones from becoming lost and shepherds them over difficult crossings in the trees.

Talapoins are further like squirrel monkeys in depending mainly on fruit and insects for their diet, and in having very precise mating and birth seasons. It seems that both the African and American forests have a niche for a small-bodied primate with a fairly omnivorous diet, and that these similarities lead to others. This is a good example of convergent evolution: similar life styles have given rise to similar creatures, although their ancestry is very different.

The talapoin is very closely related to the guenons, so much so that some authorities allot it to the genus *Cercopithecus*, rather than recognizing *Miopithecus* for a single species. However, the more that is learned about the talapoin's behaviour, the more distinct it seems to be, and it has one important physiological characteristic that the guenons lack. Female talapoins advertise their sexual receptivity with a swelling under the tail. This is not as dramatic as in the case of the mangabeys or the baboons, for example, but it certainly serves to send a message to the males. This may be important because the males and females spend so much time apart, and if the female can broadcast a message that attracts wide attention, she may encourage several potential mates to compete for her. If so, and she gets the winner, she will have effectively chosen a good father for her offspring and her male descendants will, in turn, be more likely to inherit the capacity to be sexual 'winners', thus maximizing one way of having many descendants. That, at least, is one theory to account for this characteristic of the females.

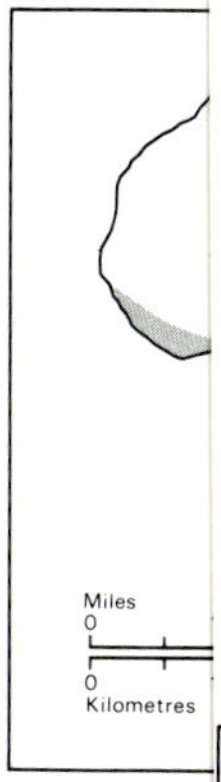

The patas monkey GENUS *Erythrocebus*

With its reddish coat and rather dashing white moustache, it is not surprising that the patas monkey is sometimes called the hussar or military monkey. The nearest thing to a greyhound that the primate

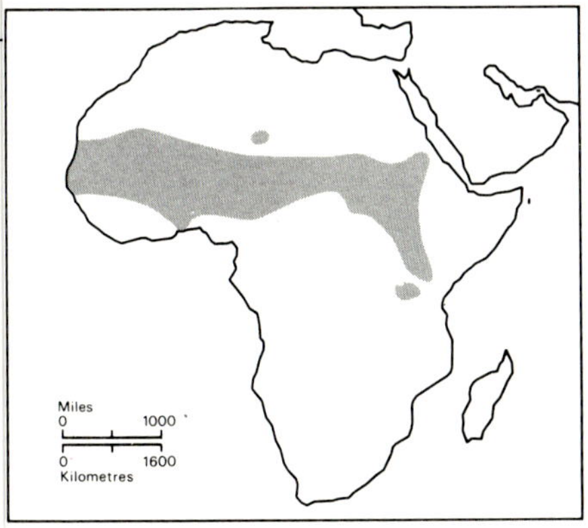

With her long limbs and athletic body, this female patas monkey is well adapted to a terrestrial way of life in some of Africa's more arid parts.

Like the other generalist monkeys of the Old World, a female grey-cheeked mangabey can stuff her cheek-pouches with food that can then be carried away and eaten at leisure elsewhere.

Male mangabeys may seem rather blasé, but they are by no means ignorant of the stage that each female in the group is at. They show an increasing tendency to inspect a female as her swelling gradually inflates, and one or more males may well begin to follow her around just before the peak, apparently keeping an eye on her. Connoisseurs all, they like to get their females at just the right moment.

Mangabeys are quite closely related to baboons, something which seems rather unlikely from a photograph but absolutely obvious the moment they are encountered in any West or Central African rain forest. They strut or lumber along big branches or sit on their haunches just like baboons, and their deep, grunting chuckles seem more appropriate for a bigger and heavier monkey. Nevertheless, they are not exactly effete themselves, and when a group is feeding, the air resounds with tearing sounds and sharp cracks as they rip bark from the trees and break off dead branches to reach the insects and spiders that are hiding in the crevices. Other invertebrates are found by rustling through the broad leaves of epiphytic plants that grow on many of the big branches high in the canopy.

Live bark is also torn off the trees; the mangabeys use their long canine teeth to get a purchase on a small part of it, then they tear it off with their hands or teeth and chew bits of it up and eat them. Large fruits add to the spectacle when hard ones as big as footballs are partially chewed before being casually allowed to fall as much as 30 m (100 ft) to the forest floor. Life below can be hazardous and

The pale eyelids of the white-collared mangabey are used as important signalling devices in social interactions.

nerve-racking. However, in spite of the general air of noisy demolition, mangabeys obtain the bulk of their diet from smaller fruit and from flowers, as is true of so many monkeys. They eat few leaves, not being equipped with the right sort of digestive system to take in large amounts. They also seem to eat little meat, but this may be a matter of lack of opportunity. Certainly, they eat birds' eggs when they find them, although some birds have been known to retaliate by mobbing individual monkeys, even to the point of harassing them so much that they fall out of the trees and cower in the undergrowth until their diminutive tormentors have flown away. Mangabeys have also been known to eat snakes, skilfully and quickly biting off the head before getting down to a leisurely meal on the body.

Grey-cheeked mangabeys, one of four different mangabey species, live in groups of about 15 or 20 individuals, in which there are usually more adult females than there are adult males. The groups have relatively large home ranges for forest monkeys: 400 hectares (about 1000 acres) is not unusual. They seem to need these big ranges because they are so effective at demolishing the food supply – especially the insects – in any one place that they then have to move on, with little incentive to come back until things have recovered enough to be worth hitting again.

When a grey-cheeked mangabey group is foraging, it spreads out on a broad front. In spite of the fact that more than one group may use the same part of the forest, two groups seldom meet during the normal course of events. They make a point of avoiding each other by moving away if they hear another group at close quarters – say less than 600 m (660 yd) away. They have a special loud call, known as a 'whoop-gobble' with which each group warns others of its location, sounding in the process rather like a rather hesitant turkey with a frog in its throat!

There are about a dozen different types of mangabeys, but they are usually classified into four multi-racial species. The grey-cheeked mangabey is one of two mainly arboreal species that only come down to the ground for short periods of time; the other is the black mangabey. Smoky and crested mangabeys, on the other hand, spend enough time on the ground to be regarded as semi-terrestrial. Possibly because mangabeys are such active, striking monkeys, many of the different races have been given their own popular names, which can be rather confusing; but the division into four species simplifies things somewhat. All weigh approximately between 7 and 11 kg (about 15 to 25 lb), the males being rather larger than the females.

Generally speaking, they are confined to the forests of West and Central Africa, although thousands of years ago they were more widespread in the east. Two small, isolated populations of crested

ABOVE Hamadryas baboons in Ethiopia. The male is yawning, probably because he is tense as a result of the approach of another male towards him and one of his females with her offspring.

LEFT The swollen rear end of the female chacma baboon signals that she is at the peak of her sexual receptivity. At this point in her menstrual cycle, she is of most interest to the males and most likely to conceive after mating.

A big male chacma baboon keeps watch on his surroundings from the top of a convenient termite mound. When a group of monkeys is feeding, one or two of them often act as lookouts from time to time. Any monkey in the group may do this, but in some species it is noticeable that it is much more likely to be a male pattern of behaviour rather than a female one.

Meat-eating occurs regularly among baboons. This big male anubis baboon is eating a freshly-killed gazelle. His companion has reason to be hopeful, as the meat is sometimes shared.

their allegiance. Most harems consist of from one to four females, but up to ten have been recorded. In contrast to the harem society of geladas, there seem to be no close bonds between the females; their togetherness is totally dictated by the power of the male.

Infants spend their early days within this rather closed society, sticking close to their mother, but definite sex differences in behaviour become apparent even in the first year of life. Although both male and female infants are groomed and nursed more or less equally by their mothers, the males show a much stronger tendency to head off for play groups with peers of their own age and older. Many of the females, of course, are adopted by their future mates before very much more time passes, while the males gradually spend more and more time together and show an increasing tendency to live on the periphery of the troop when it is at the sleeping site.

As they grow older, they become more inclined to groom each other rather than to play, and well before they are adults, they often enter into another, more permanent, kind of male–male relationship. Usually, this arises as a result of a male's attaching himself to a harem unit and being tolerated as a follower. In time, he collects his own females, but the two harems continue to move together, deciding their direction of travel on the basis of a compromise between the two males. It is evident that the older male continues to be rather 'more equal' in this decision-making, even when he becomes so old that he loses all his females. It is possible that this deference to age is adaptive for all, in so far as older animals may remember sources of food or water in times of critical shortage.

Water must sometimes be obtained by digging holes in the sand of dried-up riverbeds and waiting for the cool, clear liquid to seep in; and the baboons range over several square kilometres in search of their rather spartan diet of grasses, roots, flowers and occasional fruits. Insects are also taken, with rare swarms of locusts being a particular treat. Although they do sometimes eat meat, they do not appear to be nearly as carnivorous as their relatives that live on the African savannahs.

There are four closely related species of savannah baboon, occupying most of Africa south of the Sahara with the exceptions of areas where man has eliminated them, the area that is occupied by the hamadryas baboon, and deep inside the larger rain forest blocks. In spite of their collective name, however, they are found in some tracts of rain forest.

Typical of them is the olive baboon, the most widespread of the four, found from the extreme west of the continent right across to the Sudan and the East African countries. They are generally dark brown in colour and the adult males have something of a shoulder mane, but nothing like that of the hamadryas males. They are, however, much heavier at about 22 to 30 kg (49 to 66 lb). As among all savannah baboons, the females are about half the size of the males.

Nevertheless, the females are rather less oppressed among olive baboons than are their cousins from the arid regions. Harems do not exist and their society is based upon a mixed group of individuals, with huge variations in the numbers involved. Groups with as few as eight members and as many as 200 have been seen, but 30 or 40 would probably be more normal. Their home ranges also vary in size and overlap with each other, but the different groups show a distinct tendency towards mutual avoidance.

Social relations within the group are dictated to some extent by a loose coalition of dominant males, but these only become possessive about certain females when they are at the peak of their receptivity. For the rest of the time, the dominant males are not concerned about the females' mating behaviour. Assuming that the females conceive when they are most swollen and receptive, this makes perfect sense in terms of his dominance, permitting a male to maximize the number of offspring he has. As far as each male's evolutionary fitness goes, maximizing his number of descendants is the major aim; and among all baboons, it seems that it is the fierce competition between the males, with victory usually going to the biggest and toughest, that has led to their huge relative size and fearsome dental weaponry. Although most disputes are settled by threats and an established dominance status difference between individuals, challenges sometimes have to be made and met, and the costs for the loser may include wounds or even death.

Male baboons also gain from their size and strength in being a match for many predators. Jackals, for example, could easily threaten the smaller members of the group but, as among patas monkeys, can be instantly put to flight by any big male. The males do indeed act as protectors by counter-attacking some predators, or forming a rearguard when the group flees others. This is by no means a universal rule, however, and where one group's males might go for a cheetah or perhaps even a leopard, those of another group might flee further and faster than any of their females and young. Such a variable degree of heroism shows an obvious similarity to the human species. At night, the whole group retires onto the branches of large trees for safety.

Groups of olive and other savannah baboons are close-knit societies within which the males may be dominant, but the females are – as usual – the more permanent members. They almost always keep to the group in which they were born, whereas males transfer between groups at least once, and maybe several times, during the course of a lifetime.

The group does not split up during the day, as do bands or troops of hamadryas baboons, but forages or rests as a single unit, albeit

Among chacma baboons, as in most primates, the bond between mother and infant is a close one and crucial to the normal healthy development of the young animal. Even if a baby monkey receives plenty of warmth and food, if it lacks a mother, it will never grow up into a normal adult in terms of its behaviour.

one that may spread out on a broad front. They are generalist feeders, eating a great variety of vegetarian and invertebrate food, according to what is available. They also eat meat, commonly killing hares, birds, young antelopes and even their fellow primates in the form of bushbabies and baby vervet monkeys.

Most of these animals are pounced upon opportunistically when accidentally flushed out of their hiding places in the vegetation, but at least one Kenyan group of baboons has developed cooperative hunting techniques. It seems that they learned from accidental relay chases, and the males now drive prey, such as small gazelles,

The adult male mandrill is unmistakeable, but not as terrifying and fierce as he looks – unless he is threatened.

towards each other and some sharing of the kills has developed within the group. Before they developed cooperative hunting, these baboons never shared any of their food. In undergoing what are clearly cultural changes in the behaviour of a single group, the baboons show similarities with some macaques; and in their learning to hunt mammalian prey, there is a close parallel with the behaviour of some chimpanzees. Considerable intelligence and adaptability is evident in all these species.

The baboon genus has two more members that are rather better known in captivity than in the wild. These are the drill and the mandrill, dark-coloured animals from the forests of West Africa. The males are typically massive and probably heavier even than olive baboon males, yet the females are small by the standards of any baboon species. Both sexes, however, have relatively huge muzzles with prominent swellings running lengthwise on either side of the nose.

Drills and mandrills are justly renowned for their gaily coloured rear ends which have hues that range from deep scarlet and various reds and pinks to blue and a delicate lilac. It has been suggested that the significance of the overall effect lies in its virtual luminosity in the dark blue-green light of the forest, because it allows other individuals to follow the males. As a variation on the Pied Piper theme, this idea is quite enchanting, but somehow unconvincing because forest monkeys are usually well able to locate each other, however dark their habitat and their bottoms. It seems more likely that big male drills and mandrills are emphasizing their social and sexual positions by means of permanent colour signals. There may be a parallel here with the blue scrotums that so clearly identify male patas monkeys and some male guenons.

The mandrill goes one further than the drill and more or less replicates his red penis and lilac scrotum in the colour patterning of his face. He has a bright red nose with blue, ridged swellings on either side. This is also assumed to be a sexual signal – one that is clearly transmitted with every look that he gives towards others. From what little is known about these two species, it seems that the adult females greatly outnumber the males in their groups, something that is suggestive of fierce competition between the males for breeding positions. If so, this would be in keeping with baboon behaviour in general and would explain why the males are more than twice as big as the females and why they signal their status in such a dramatic fashion.

The gelada GENUS *Theropithecus*

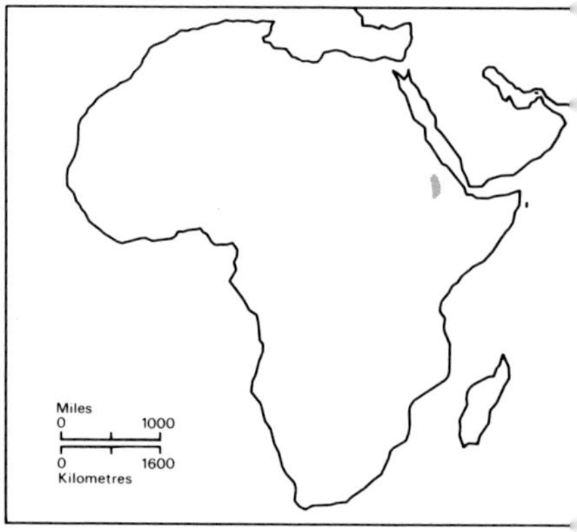

The high Amhara plateau of northern Ethiopia is a cool, windswept land that rises in places to 4500 m (nearly 15,000 ft) above sea level and is cut by spectacular gorges, often hundreds of metres deep. That of the Blue Nile, flowing out of Lake Tana, in places flows 1500 m (nearly 5000 ft) below the surrounding highlands. Monkeys are not known to suffer from vertigo, but there is still something rather breathtaking about a primate that retreats over the rims of these canyons to find refuge by clinging to rocks and bushes above such a sickening drop. So attached are the geladas to their vertical sleeping sites that they are never found more than 2 km (about $1\frac{1}{4}$ miles) away from them.

They are spectacular, baboon-like monkeys with males that weigh about 20.5 kg (45 lb) and females about 13.5 kg (30 lb). The males add to the difference by sporting a heavy, leonine brown mane of long hair. Both sexes have a large, pink patch of naked skin on the chest, which in females is lined with a necklace of fleshy caruncles. These swell up when she is sexually receptive.

There were once at least four species of geladas spread over eastern and southern Africa. They specialized in a terrestrial way of life and a diet of grass blades, roots and seeds. They were probably the only monkeys that have ever concentrated on such a grassy diet and they seem to have been very successful at it until comparatively recently. Nobody knows exactly when they disappeared but it may have been about 50,000 years ago; they were probably hastened away by an expanding human population that found heavy, ground-dwelling primates to be easy meat. Condemning his fellow creatures to extinction is not a new human pastime.

Today's geladas are a single species, living only in their mountain fastness. Large herds of them may still be found occasionally, even up to 600 strong, but usually these big aggregations break up into bands of 30 to 250 animals, within which there are harem groups and gangs of bachelor males. The nucleus of gelada society is the group of adult females that forms a harem. In contrast with the hamadryas baboons, which have a superficially similar arrangement, gelada females are not herded by their males. In fact, their male 'lord' has precious little scope for dictating to them. The females stay together as a group because of the bonds between them; and a young male's aim in life must be to leave the bachelor society and gain a group by competing with other males. He may do this by fighting with, and defeating, a resident harem male, or he can tag along after a harem as a follower and try to build up friendships with some of its younger females. If all goes well, he may either inherit all the females, or the younger ones he has so assiduously courted may split off as a group to form a new harem with him in attendance.

Bands of geladas usually start their day in the cool of the early morning by climbing up out of the gorge where they spent the night and embarking on their regular routine of warming up with a round of social activity: grooming, quarrelling and chasing, and so on. By about nine or ten o'clock, they usually start to get hungry and they literally sit down to eat. Some 90 to 95 per cent of their diet consists of grass, which they pull and dig out of the earth, chewing up the roots as well as the rest of the plant, before shuffling forward on their bottoms to have a go at a new patch. The procedure is continuous, with the gelada chewing up one handful as it collects a second.

Geladas have highly opposable thumbs that make them extremely efficient at pulling up grass. They are so successful at it that they maintain much higher population densities, 70 to 80 per km^2 (about 195 per square mile), than the similarly sized baboons of the genus *Papio*. They top up their grassy diet with very small amounts of the leaves, fruits and flowers of herbs and bushes, and even an insect once in a while. They seldom climb trees for any reason, being rather inept and clumsy when they do so.

OPPOSITE **The pink chest patch is unique to the gelada. When a female is in the sexually receptive part of her cycle, fluid-filled vescicles form a necklace of beads around the patch. Some also form on her rear end. As they are only present for a short time, they give a very definite signal about her condition.**

RIGHT **The social bonds are close within a gelada harem group, particularly between the females. These individuals are carefully grooming each other's hair as they sit in a row. Hamadryas baboons have a superficially similar society, but the females do not form close bonds with each other.**

Recent estimates suggest that there are rather more than half a million geladas left in the world today, but they are coming into greater and greater conflict with local farmers, who are cultivating increasingly steep hillsides in their habitat. And, in the southern part of the Amhara plateau, the local tribesmen slaughter the big males every eight years when custom demands that they use the manes in their traditional regalia at coming-of-age ceremonies. So far, the geladas have survived this unfortunate appreciation of their finery, but wiping out most of the adult males at regular intervals must be socially disruptive, to say the least. It would be a pity if this extraordinary animal were to be harassed into oblivion. It has, after all, been shuffling around Ethiopia on its bottom for thousands and thousands of years, which must give it some sort of squatters' rights.

The macaques GENUS *Macaca*

For macaques distribution map (see overleaf).

About halfway up the Rock of Gibraltar, at a spot that is regularly visited by tourists, some extremely fat monkeys relax in what little shade there is available. They do not have to worry about going off up or down the cliff to forage amid the sparse vegetation; they have servants to bring their food. These are the famous 'rock apes' of Gibraltar, and the British Army (plus an informal army of tourists) sees to it that they are well fed. Not so many years ago, they were kept on short rations and took to raiding houses and gardens in the town below, to the considerable anger of the human residents. Now a constant supply of food anchors them in place: one group at the 'Apes' Den' halfway up the slope and another on the very peak of the Rock.

The 'rock apes' are not actually apes at all; they are Barbary

up whom if it comes to a fight; and none of the dominant males will come to the defence of an adolescent male.

Adult females do fairly well, since they are not picked on as much as others by the big males and they can hold their own against most of the younger animals. It is not surprising that the males are generally (but not entirely) pleasant to them, since the females have obvious attractions in the mating season. (Unlike many other macaque species, toques have quite rigid breeding seasons.) The class that does worst in terms of being thoroughly browbeaten are the very little females. The first two or three years of their lives must be extremely unpleasant, with all the other monkeys being nasty to them and their being particularly nasty to each other.

The harassment the females suffer when very young and the males are subjected to just as they are approaching adulthood causes them almost constant stress and deprives many of them of the good food they need to remain healthy. It has been calculated that 90 per cent of males and 85 per cent of females die before they grow up, and the most likely age at which to die is that at which they are most harassed. Thus, the females tend to die five or six years before the males. Nobody knows whether they die of stress, starvation or disease, but a combination of all three seems most likely, with fighting as a serious additional problem for the up and coming males.

During the breeding season, individual females start to become sexually receptive and the males compete for their favours. Vicious fighting often takes place and males that are low in the hierarchy

ABOVE LEFT **Over-feeding has made some of Gibralter's Barbary macaques fat and lazy. Here, a female is grooming a big male.**

ABOVE RIGHT **The toque macaque is found only on the island of Sri Lanka. This young male has been disturbed in the act of drinking by dipping his fingers into water that has collected in a tree-hole.**

of one group often move out to test their fortune in another. The successful males go off with the females so that, where there are many toque macaques, the woods tend to be full of these couples, repeatedly copulating between meals and rests. For the losing males, this is the most likely time to die.

Once a toque macaque survives beyond the dangerous period in his or her life, there is every chance that he or she will live to a middle age in the late teens or even senility in the late twenties. The whole set-up is thus an excellent example of the way in which far more individuals are born than can possibly survive, and the winners are selected from the losers on the basis of fierce competition. Monkeys that are sickly or weak or stupid enough to pick a fight that can only end in disaster are consigned to the evolutionary dustbin. Those that are healthy and strong and time their challenges well are the 'fittest' ones that survive and leave descendants.

The system reaches an overall equilibrium and, because ultimately much of the competition centres around the food that is available, the total number of survivors matches the food supply. In times of scarcity, more individuals die; in times of plenty, more survive. What is *not* happening, however, is some sort of group or species consensus about how many toque macaques can live. The total population stays at whatever number the environment can support solely as a result of the constant competition between individuals. Once again, animals do not behave 'for the good of the species'.

Among Japanese macaques, mating takes place with the male fully mounted on the female, clasping her legs just above the ankle with his feet.

All known macaque species have dominance hierarchies in their social systems, but the Barbary macaque has added a neat twist to the art of social manipulation. These monkeys are rather nicer to each other than are the toque macaques, and the males are considerate towards their young, often giving them rides on their backs, helping them over rough terrain or saving them from the attacks of dogs or jackals. It seems that the males are so well disposed towards the infants that the involvement of an infant in an encounter between two males is enough to ensure that the dominant one will not attack the subordinate. This has given rise to a system in which males frequently 'borrow' babies to present to other males. The presenter is usually the subordinate of the two, and it seems that, in so doing, he manipulates the receiver into being friendly enough not to drive him away. One cannot imagine the ploy working for toque macaques, especially if the presented infant were one of the persecuted little females!

Males of the Japanese macaque also seem to be particularly caring towards the juvenile members of their group, but subordinates do not appease dominants in the North African way. Japanese macaques belong to a species that occurs further north than any other non-human primate. They reach 41 °N at the tip of Honshu

Island, where great snowdrifts pile up in the winter and no leaves grow for nearly half the year. During this barren period, these stocky, thick-furred animals somehow manage to keep warm and they survive largely by eating bark.

There is, however, one group that lives in a mountainous area in central Honshu and has hit upon a novel way of coping with the cold. On 16 February 1963, a two-year-old female, called by her observers Mukubili, took the plunge into a local hot spring. There was snow on the ground but the temperature in the pool was about 43°C (109°F). She was quickly followed by a young male and it was not long before other youngsters took up the habit. In time, it became established as a purely winter pastime of many monkeys. They test the water first by means of a few tentative splashes with the hands and feet, and then they ease themselves into the delicious warmth. Once there, they swim about like dogs, plunge under the water and even rub behind their ears. It is, however, most important that they get out of the water in time to dry off properly before night falls in order to avoid freezing to death where they sleep.

Mukubili is not the only genius among Japanese macaques, and, indeed, she is rather overshadowed by the inventiveness of Imo, a female who came up with two clever ideas on the little island of Koshima in the extreme south of Japan. Imo was born in 1952, the year in which biologists who wished to study her group began to give the monkeys food on a regular basis. The monkeys received frequent supplies of sweet potatoes in this way, but as they were distributed on the beach, they were always covered with sand. They rubbed this off as best they could with their hands, but the sweet potatoes must still have been rather gritty to eat.

Imo solved the problem in September 1953, when she was only about 16 months old. She dipped a potato into a stream that ran across the beach and brushed off the sand under water with her other hand. And presto – the result was a clean potato! It took a while for the idea to spread, but by early 1958, 2 of the 11 adults in the group and 15 of the 19 individuals aged between two and seven had learned the trick. By 1962, all but the older adults were potato-washers. The habit had spread through the group, with Imo's playmates of her own age, and her mother, being the first to learn. Apart from Imo's mother and one other female, none of the monkeys who were three years old or more in 1953 ever did learn to wash their potatoes, which just goes to prove the adage that you cannot teach an old dog new tricks.

Among humans, one invention often leads to another, and so it has been with the Japanese macaques. Not satisfied with clean potatoes, many of them took up salting them and by about 1960, potatoes were being dipped in the sea as often as in the freshwater stream. Monkeys would pick up a load of potatoes and carry them to the sea, wade in, wash them, and dip them in the salty water between bites. In time, all the practice that they got improved the monkey's ability to walk on two legs and their ability to swim. From being a group of normally quadrupedal macaques that seldom entered the sea, they became an extraordinary group of macaques that would grab a load of potatoes, perhaps wash some in fresh water, or perhaps walk bipedally for 50 m or more, before wading bipedally into the sea to dip and eat. The infants soon excelled at swimming because they were regularly carried into the sea by their mothers at a very early age.

For a few years, the monkeys had an additional problem brought about by the inclusion of wheat grains on the menu. These had to be picked out of the sand one by one and were eaten in a thoroughly gritty condition. In 1956, at the age of four, Imo showed that her earlier genius had not been a mere flash in the pan. She picked up a handful of grains and sand and threw them into a pool of water. The sand naturally sank, the wheat floated, and Imo scooped out her clean food.

Once again the new habit spread through the group via Imo's associates, but by this time she was older and socializing more with monkeys to which she was related through her mother. Female Japanese macaques are unlikely to have any special relationships through the paternal line since mating is fairly promiscuous, but bonds through the maternal line are strong and result in distinct clustering within the group. Thus, it is not surprising that it was mainly through the 'matriline' that wheat-washing spread. Once again, many of the old individuals never learned the new pattern.

Imo's inventions show that radical thinking is adaptive in the face of new circumstances and perhaps also that youth may have something to offer. Unfortunately, what happened next shows that exploitation can happen even in the relatively simple society of the Japanese macaque. Some monkeys learned to wait for others' grains to float towards them, a sensible enough use of left-overs, but others learned to wait until the wheat-washer had thrown the mixture of wheat and sand into the water, whereupon they leapt in and snatched the food!

Perhaps the most important lesson to be learned from the Japanese macaques is that at least some monkeys have the capacity to generate new ideas and spread them by means of cultural propagation in an historical sequence. Another group of Japanese macaques has shown how new ideas can be suppressed by those in authority: a dominant male attacked members of his group who accepted fruit from humans, and it was not long before no monkey would look at a hand-out even though they would raid nearby gardens for exactly the same fruit.

The macaques are noteworthy not only for their inventiveness

and their fascinating societies, but also for the contribution they have been forced to make to medical science. During the late 1950s, India alone used to export well over 200,000 rhesus macaques *every year* to the laboratories of the Western world. Most of these monkeys were used in the development and testing of vaccines against polio and other diseases but, among the remainder, many were used casually and needlessly for senseless or dubious experiments. In 1977, Indian macaques were saved from further large-scale persecution when the Indian government banned their export. This caused something of a panic among laboratory scientists and health authorities of the West because they had become used to relying on captured wild animals for their needs and had not done enough about breeding the monkeys in captivity. However, another macaque was available to have its services 'volunteered'. This was the long-tailed macaque of South East Asia, which is now flown out to the world's laboratories on a regular basis, although, happily, new research techniques mean that the years of peak demand have passed and monkeys are only traded now at a fraction of the level that prevailed two decades ago.

ABOVE LEFT **The pig-tailed macaque of South East Asia's rainforests show many similarities with Africa's forest baboons. The males are much bigger than the females (this is a male) and groups may travel over large areas in search of food, moving long distances on the forest floor but foraging at all levels above the ground in the trees.**

ABOVE RIGHT **The long-tailed macaque is found beside rivers and coastlines, and in farmland, throughout much of South East Asia. Every year, thousands of these animals are sent to biomedical laboratories, mainly for use in the production and testing of drugs and vaccines.**

OPPOSITE **Orang-utans learn to climb in the trees from a very early age. This infant belongs to the Sumatran race (see page 184).**

6. THE TAILLESS PRIMATES: APES AND MEN

This last group of primates includes the biggest, the brightest, the strongest, the fastest, and the most widespread members of the order. Man commonly thinks of this group as being the 'highest' – but that is just because he likes to place himself at the top and 'highest' comes to mean 'most like me'.

There is no universal agreement among biologists as to how closely related man and the various apes are to each other. Some argue that humans diverged from the ancestral stock more than 35 million years ago, and have no ape ancestry because the apes themselves had not yet become apes. However, as more and more evidence becomes available, this point of view looks less tenable.

At the present time, the most popular view is that man and the apes took different evolutionary courses about 20 or 25 million years ago, but even this seems like a long time when some of the startling similarities are considered. The plain fact is that when the molecules that make up the various species are examined, they show man, the chimpanzee and the gorilla as being exceedingly closely related to one another. Biochemically, these three genera are identical in the *majority* of their composition. The chimpanzee and the gorilla are, of course, the African apes. The orang-utan and the gibbons, both from Asia, form another closely related unit. All monkeys are much further away in biochemical terms. To summarize, the biochemical evidence suggests that the ancestral ape stock diverged into African and Asian groups, and that humans did not break away from the ape line until quite some time later.

If this view ever gains really wide acceptance, it will become necessary to place man and the African apes in one family, with the Asian apes in another. However, the biochemical evidence is not all that must be considered, and currently accepted taxonomy reflects more than anything the degrees of anatomical similarity that are found between the species. The small, long-limbed gibbons, or lesser apes, are placed in the family Hylobatidae; the large-jawed, powerful, great apes are placed in the family Pongidae; and the small-jawed, rather lithe, upright, large-brained human is placed in the family Hominidae.

Actually, when looking at the many differences between, say, a diminutive talapoin and a massive baboon, both of which are members of the same subfamily, the anatomical justifications for placing humans and chimpanzees in different families do not seem terribly convincing by comparison. There is, however, the behavioural evidence to consider. There can be little doubt that the human species, with its culture, history and language, is way out in front of the apes on these scores, and yet the more we learn about other primates, the more obvious it becomes that the differences are matters of degree rather than matters of absolute presence or absence of classes of abilities.

Humans were once thought to be the only primates that could use tools, but that idea was disproved more than 50 years ago when chimpanzees in Cameroon were seen poking long twigs into underground bees' nests to extract honey. Since then, it has been discovered that chimps actually make tools by fashioning twigs to their needs.

Then humans could be defined as the only primates with history and culture, but a quick look at the macaques of Japan, or at the Kenyan baboons that have learned to hunt, rapidly dismisses that. If only we had more evidence, we might find that the green monkeys that have moved into the farmland of Bakossi are in the same category; and it seems very unlikely that there are not numerous other examples.

Language was perhaps the last bastion in the possibilities for an absolute distinction for man, yet even that has now fallen. True, no ape or monkey can talk or write, but what is so special about these modalities when apes can use a sign language or punch out syntax on a computer?

The more we learn about non-human primates in general, and the great apes in particular, the more we find that we have underestimated them; and the more correct it seems to consider these last five genera as a fairly cohesive group that has more to unite than to divide it.

I The Anthropoid Apes: acrobats and singers, intellectuals and strongmen

The term anthropoid simply means 'man-like' and is a reflection of the observation that apes are in many ways more like people than they are like monkeys. In at least one respect, they outdo our own species, since aberrant individuals with small tails are apparently even rarer among apes than among humans.

Most of the differences between apes and monkeys are the result of major trends: towards even greater reliance on vision in place of smell; towards greater size; towards a more upright bodily posture; towards having fewer offspring, which take longer to grow up; and towards living longer.

All of the apes see extremely well and have less pronounced muzzles than do most monkeys, even though they retain relatively powerful jaws. As a result, they have very 'human' faces; this is especially true of the three great ape genera.

Apes and monkeys also differ significantly in the shape of the protuberances on the grinding surfaces of their cheek teeth. The apes have conical lumps where the monkeys have ridges – a division that may seem like a rather abstruse technicality, but which is indicative of the distance in relatedness between the two groups.

The tendency towards a more upright body posture among the apes, mainly because they often hang or swing under branches, has resulted in the development of a pelvis that supports sheets of muscles which in turn support the weight of the organs in the lower part of the body cavity. Because the apes (and humans) spend so much time in an upright position, the lower end of the digestive tract, the bladder and the womb all have to be held in place more from the bottom end than by muscles that stop them from falling out of the front of the body, as is most important for more quadrupedal primates. Of course, the organs do not fall about in the body cavity whichever way up the individual chooses to be, but muscles and skeletons tend to be most developed where they do most work. Hence, ape skeletons can be recognized by their supportive pelvises.

Among all of the apes, females tend to have longer intervals between giving birth than do monkeys. This is most extreme among the great apes, possibly averaging five years in orang-utans. Gestation periods are longer than in monkeys, ranging from about 210 days among gibbons to about 265 days (the same as humans) among gorillas. The comparable figures for orang-utans are 233 days and for chimpanzees 238 days. In spite of the length of time that ape babies spend developing inside their mothers, they are born in a less developed state than monkey babies, and their mothers have to be much more careful to hold them lest they fall. They also take longer to grow up (anything from 7 to 13 years, depending on the species and sex).

Thus, all of the apes have long periods of immaturity in which to learn about the world; and they follow these with long lives in which to put that learning to use. This has enabled them to develop their intelligence to a remarkable degree, but it has cost them the ability to multiply quickly. For all their cleverness, therefore, they are exceedingly vulnerable to extinction as a result of human activity, because their numbers cannot recover quickly from catastrophes. And our species does have a knack for destroying them.

The gibbons GENUS *Hylobates*

If there is one sound that characterizes the Malayan rain forest and evokes associations of steamy jungles in the early morning, it is the haunting duet of the lar gibbon. A few, brief, melancholy whoops from either the male or the female begin the song, and then the

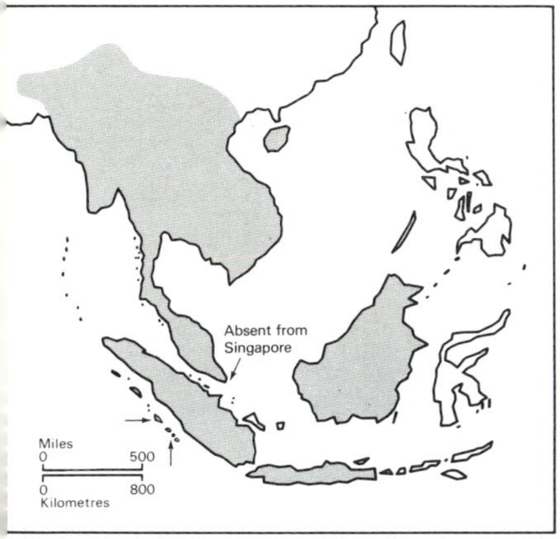

female takes over alone with a 'great call' consisting of a few small piping sounds followed by a crescendo of steeply rising notes, perhaps six or ten, that get higher and louder until they reach a peak and are followed by two plaintive little wails that die away, and the male chips in again with more of his brief, melancholy calls. For the next few minutes, the male and, to a lesser extent, the female may hoot gently, before beginning their stylized song once again.

It seems that each pair of gibbons practises singing together until they hit upon a personal variation of the species' basic song. Once they have done so, this becomes their longest duet and one by which each pair may individually be recognized, although much of their calling will continue to be incomplete versions of the song.

The lar, or white-handed, gibbon is one of the best known of the nine species of this genus, all of which are extremely agile, small apes that live in close, and highly territorial, family groups. All of them have very thick, shaggy fur and dark, bare faces. They have

An adolescent lar gibbon. Gibbons are the most acrobatic of all the primates.

callosities on their bottoms to facilitate sitting comfortably in the trees; and relatively long, slender legs and feet. It is, however, their arms that are most exaggerated, being relatively the longest of all primates and ending in long, slender hands that make superb hooks for swinging on branches.

Gibbons can run bipedally along broad branches and climb upwards with all four limbs, but it is at arm-swinging beneath branches, known to biologists as 'brachiation', that they really excel. When not resting, they spend the vast majority of their time suspended beneath branches, either feeding or travelling. When feeding, they frequently hang by just one arm and move right out to the thin, terminal branches of big trees where the choicest fruits and young leaves may be obtained. At these times, their legs dangle below them, but when they start to move fast, they tuck them up close to the body, out of the way.

Nobody has ever measured the speed of a brachiating gibbon in full flight for the simple reason that it is impossible either to keep up on foot or to drive a vehicle through the forest. There can be little doubt, however, that they are capable of reaching some truly colossal speeds, especially when whipping through the forest canopy on a steep, downhill incline. Under such circumstances, they hurtle through the air for as much as 15 m (about 50 ft) between supports and seem barely to tap the branches as they flash by. There may be faster land mammals in open country, but the gibbons must have the record for speed in the forest and it is hard to think of any way in which an animal could move faster through the trees without actually flying.

Most gibbons, including the lar, weigh about 5 to 7 kg (11 to 15.5 lb); the females are approximately the same size as the males. There is also one rather larger species, the siamang, which weighs about 10 to 12 kg (22 to 26.5 lb). Almost all of the species replace each other geographically and consequently do not share any of their habitat with each other, except in some small areas of overlap. The exception is again the siamang, which shares much of its forest home in Malaya and Sumatra with either lar or agile gibbons. The reason for this is ecological: the smaller gibbons have similar life styles and therefore compete with each other if they live in the same area, which must presumably result in the demise of one or other species. On the other hand, the larger siamang is differently adapted to its environment and may share a forest area with another gibbon species without there being fierce competition between the two.

In the Krau Game Reserve in the centre of the Malay Peninsula, lar and siamang gibbons occupy a well-studied patch of lowland forest. The lars vary in colour from dark chocolate to creamy blond, or they may be parti-coloured with dark and light patches. For example, the limbs may be darker than the body. They are always consistent, however, in having white fur on the hands and feet and a white ring around the face. Siamangs, on the other hand, are all black.

Both species live in monogamous family groups in each of which the mated pair is accompanied by up to four offspring. Wild gibbons seem to live for about 20 to 30 years, and as far as anybody can tell, they normally pair for life. Nobody knows exactly what mechanism causes them to be so faithful, but it is certainly not a purely sexual attraction since the young are born at intervals of two or three years and the adults are only sexually active for a few months when it is time to start a new baby. Thus, by means of long periods of abstinence, they practise the most basic (but effective) form of family planning.

The young take seven or eight years to become mature enough to strike out on their own, so a gibbon's family usually contains offspring of several different ages. Only one baby is born at a time, and it spends the first year of its life being closely guarded by its mother. Among the lars, as in others of the smaller gibbon species, the mother continues to be the main protector throughout the infant's second year, but among the siamangs the father takes over at this stage. He often carries the youngster, particularly when the family are moving too fast for it to keep up by itself. Over the next few years, the immatures of both species gradually become increasingly independent until at the age of six or seven, having attained their full adult size, they usually start to show signs of detaching themselves from the rest of the family. They may move a little apart from the others and sleep apart. The process appears to be hastened by extremely aggressive hints from the parent of the same sex, until the young adult ape sets off on its own and begins the business of finding a mate and a territory of its own.

In the Krau Game Reserve, the lar gibbons defend a territory of about 54 hectares (133 acres), whereas the siamangs defend one of 48 hectares (119 acres). This defence only applies to members of the same species – the lar and siamang territories overlap – and is achieved firstly by the advertisement of occupancy and secondly by actual conflict. The advertisements are the duets of the adult pair, which not only inform rivals that the territory is still occupied today, just as it was yesterday, but also make the individual identity of the animals and the cooperative nature of their relationship very clear. During some periods they may sing several times a day, yet at other times, weeks may pass with hardly a song. On average, the lar gibbons sing their melodies about once a day, usually between 7 and 9 in the morning. The siamangs sing about one-third as often, and time most of their calls between 8 and 10. Possibly, by calling earlier, the lar gibbons avoid being drowned out by the siamangs, for although the lars are superior at melody, the siamangs have the

The siamang is the largest of the gibbons. This young Sumatran female shows obvious similarities in her long limbs with the spider monkey shown on page 111. Both species are specialized for swinging under branches and, although they are not at all closely related, convergent evolution has made them rather alike.

edge on volume. The male of the latter species gives resonating booms but is put to shame by the female who gives an extraordinary series of barks and booms that sound like two callers at once. The male then signs off with a scream. As with lars, each pair works out and maintains a particular pattern. Both species may be heard for at least a kilometre in the right conditions, the siamang possibly a little further.

The actual conflict in defence of the territory takes place when groups of the same species approach their mutual boundary. The males may then spend a few minutes, or even an hour or two, alternately chasing each other and displaying their acrobatic skills, while the females wait in the background, call occasionally and sometimes groom their mates during pauses in the action. Obviously, the frequency with which these conflicts occur depends to a great extent on the number of neighbours that a gibbon group has; but in any case territorial defence of one sort or another consumes a great deal of a gibbon's time and energy.

This illustrates the significance of their territories to these animals. Since they are prohibited from foraging outside them by their neighbours, they must be big enough to provide food and trees for sleeping for all the family. Importantly, they must provide enough food to cover the most critical periods of shortage, not just enough for average times.

Thus far, the siamangs and the lars have seemed remarkably alike, but they are distinguished by the way in which they use their territories and the food they harvest. On an average day, the lars travel about 1.5 km (1640 yd) around their territory, whereas the larger siamangs only manage about half that distance. Furthermore, the lar family spreads out more than that of the siamang,

The slender hands of the lar gibbon make ideal hooks for hanging over branches with the body suspended below.

and the lars visit more discrete food sources than do their larger relatives.

What is happening is that the lars are concentrating on fruits, many of which are rather dispersed in the canopy and can be reached only if the gibbons move right out onto the smallest and springiest branches. Being smaller than the siamangs, they seem to be better at this. The siamangs also eat a lot of fruit, but they eat more foliage than the lars, much of which comes from the biggest trees in the forest. As a consequence, they can feed for longer in one place. Presumably, the lars also gain more energy from their fruit-rich diet than do the siamangs, so it is not surprising that they move about more.

Both species are particularly fond of figs, many different kinds of which grow in their forest. (The common, cultivated fig has a huge number of wild relatives.) They also eat flowers, insects and spiders, and avoid any plant material that is loaded with toxins (poisons), so there is considerable overlap in their diets. What distinguishes them, and enables them to share the same habitat, is really quite a subtle difference between the two. The lars use more energy looking for more widely dispersed, rich food, whereas the siamangs are more inclined towards energy conservation.

The reason why there are so many small, similar gibbon species living in South East Asia – those other than the siamang – seems to be that they had a common origin about two million years ago; but, since then, climatic fluctuations and changes in the sea level have at times isolated populations, and at other times allowed them to come back together again. During some periods, the Malay Peninsula, Borneo, Sumatra, Java and some smaller islands were united into a huge subcontinent known as Sundaland. When it was covered with one vast forest, which was not all the time, the gibbons could move around freely; but sometimes great swathes of grassland intervened and, with the ending of each ice age in the northern hemisphere, the water that was unlocked from the great ice sheets caused the sea to rise and break up Sundaland into its various islands. When another ice age came, the sea level dropped again and Sundaland re-formed. During his travels in the area more than a hundred years ago, Alfred Russel Wallace noted that the degree of similarity between the fauna of any two islands depended far more on the depth of water between them than on how close they were to each other. The reason is obviously that it takes a bigger drop in the sea level to make a land bridge out of a deep channel than a shallow one, whatever the width.

In order to work out why each gibbon species is where it is, an enormous amount of reconstruction has to be done. Since ocean, river and grassland barriers have to be accounted for, the topic is horribly complicated and by no means resolved. The present position includes such anomalies as agile gibbons in parts of Borneo and Sumatra, lar gibbons in parts of Sumatra, Malaya and Thailand, and yet another small enclave of agile gibbons with lars on either side of it in northern Malaya.

It also seems to be the case that when some gibbons are reunited after having been separated for thousands of years, they may not have evolved sufficient differences to prevent interbreeding (hybridization) between the two. Hybrid animals have been seen in both Thailand and Malaya, but we do not yet know how much of this is due to man's influence on the habitat – perhaps influencing the behaviour and distribution of the gibbons. Nor do we know what happens to the descendants of these hybrids. Does natural selection eventually eliminate them as misfits? We may never know the answer because wild gibbons cannot survive without large tracts of relatively undisturbed, tall forest in which to live, and their home is disappearing at such a speed that only a few isolated pockets will still exist at the end of this century, if present trends continue. Thus, even the fittest among them will leave few, if any, descendants.

The orang-utan GENUS *Pongo*

Asia's lesser apes may be among the fastest of all land animals, but the continent's only great ape normally makes a point of proceeding as one of the slowest. Orang-utans can speed up when suitably motivated – panicked individuals may even leap across small gaps in the trees – but for most of their travel they proceed in a deliberate, quadrumanual fashion. It seems only right to describe them as having four hands, rather than four feet, because even when they

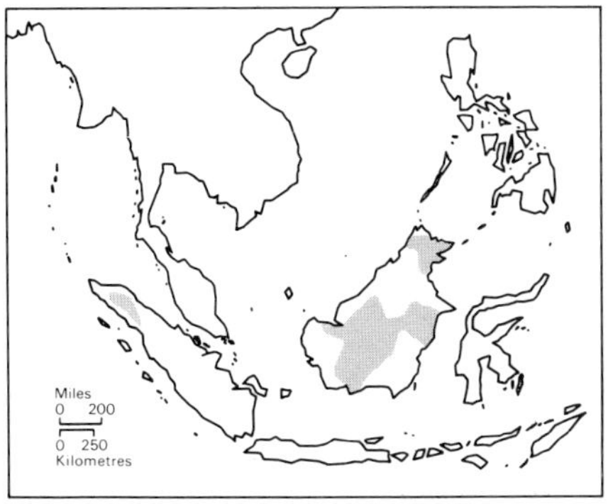

use all four to move along a horizontal branch, they tend to grip the support rather than stand on it, unless it is too broad.

They move their arms and legs in whatever order circumstances dictate, being equally at home hanging sloth-like under a branch or spreading their weight over several slender supports. They quite often swing under branches, but seldom proceed for long by means of true, arm-swinging brachiation. To cross gaps between the trees, they reach out, catch whatever they can of the next tree, and haul it in until they can grasp a firm enough support to transfer their weight. Such gaps are difficult for the little ones to cross, even when they can move confidently within a mass of branches, so their mothers hold the gaps together for the youngsters to scramble across, or make bridges out of their maternal bodies. Scrambling across mother's body, up to 40 m (130 ft) above the ground, is routine for a two- or three-year-old orang.

Compared with the other apes, orangs are fairly unsociable creatures, and the longest bond between individuals is that between mother and offspring. The young stay with their mother until adolescence at about the age of seven, but after that they generally only see each other from time to time. Females grow up quicker than males, reaching their adult weight of about 35 to 50 kg (77 to 110 lb) at the end of their eighth year. Males go through about five years of a 'subadult' phase from 10 to 15 years, and only after that do they become physically and socially mature, with weights of between 45 and 100 kg (99 to 220 lb).

When a male goes from the subadult to the adult stage of his life, he rapidly puts on weight and develops great, fleshy flanges on the sides of his face, a high, fatty crown to his head, and his hair becomes longer and darker. In Sumatra, the males also grow long beards at this time. In a way that is not properly understood, the full maturation of subadults is sometimes delayed by the presence of older, dominant males. Presumably, the fact of being dominated has an effect on the hormones that trigger full maturation.

Orangs come from the South East Asian islands of Borneo and Sumatra, and are only found in restricted areas even there. To the expert eye, however, individuals from the two islands are quite distinct and form two separate races. To the layman, on the other hand, they are remarkably alike, although the epitome of the Sumatran race is longer-faced with lighter, longer hair than his Bornean counterpart.

As among all the apes, the interval between births is long: $4\frac{1}{2}$ to 5 years is normal for a female orang. Adult females usually spend their days travelling slowly around a range that must be quite familiar to them, accompanied by one or two offspring. Like all members of their species, they show a remarkable ability to arrive at certain trees just as their fruit is ripening, even though the Asian

For illustration of an infant orang-utan of the Sumatran race (see page 175).

OPPOSITE **An adult male Bornean orang-utan. His cheek flanges will probably get bigger as he gets older and he may develop a fleshy crown on the top of his head. However, it is generally only among fat zoo animals that these characteristics get really exaggerated.**

RIGHT **The feet of an urang-utan (centre) are adapted for grasping branches and other objects. They make a very effective second pair of hands.**

rain forests are among the most complicated in the world. One of the uses to which orangs put their considerable brains must be to sorting out the thousands of species of trees in their area and remembering when individual trees will fruit or flower, even though they do so in irregular cycles that are often quite unfathomable to the human observer.

Several orangs will often meet at a fruiting tree, especially if it is a big fig, but they usually show little interest in each other. Youngsters might take the opportunity to play with each other, since for most of the time they have to be content with solitary play. If a male arrives, the situation sometimes changes radically. Big males can precipitate the flight of all the other orangs as soon as they come into view; yet at other times males and females studiously ignore each other. Females do not indicate their degree of sexual receptivity by means of an external swelling, so it is not surprising that males sometimes attempt to check them out with a close and very personal inspection before getting back to the more important business of eating.

As likely as not, any aggregation of orangs that meet at a fruit source will be quite mixed in terms of where they come from. Some will be resident animals that know the other locals extremely well from frequent encounters, whereas others are likely to be migrants, although they may have passed through the area before.

Orangs have home ranges that can be identified in terms of where they spend long periods of time, but there is nothing to prevent them from leaving for new territory, if it suits them. In both Borneo and Sumatra, it has long been known that orangs tend to congregate in certain areas when fruit is most abundant, often on a roughly seasonal basis. Yet the societies are different on the two great islands. Bornean females tend to have solitary home ranges of roughly 100 hectares (about 250 acres) and meet each other according to the results of their individual decisions about where they are going on any one day. Sumatran females, on the other hand, move in wider areas, perhaps 5 km^2 (about 2 square miles), and give the impression that they are grouped in very loose communities which coordinate their meetings.

Orangs of both races may occasionally team up into travelling bands of maybe six or a dozen individuals that wander through an area, perhaps with little apparent contact between different individuals or subgroups of a band, but nevertheless all moving slowly along roughly the same line of travel. Such a band may well appear to be centred around a big adult male who gives long, loud calls at frequent intervals.

Adult males have a call that is as spectacular as their appearance. It starts with a series of rather low-pitched bubbling sounds and gradually rises through loud groans into a long, powerful, gravelly roar that carries for over a kilometre. These 'long calls' last for up to a minute in Sumatra and three times that in Borneo, which must be a considerable effort, considering the volume, even though their production is aided by a large laryngeal sac that gives the males the appearance of having a huge double chin. Subadult males do not call, and neither do all adults. The habit seems to be largely, but not entirely, restricted to dominant individuals, particularly those that reside more or less permanently in one area, and those in the travelling bands. Quite a few males – perhaps even the majority – travel around nomadically on their own, and they are rather less likely to give long calls.

Nevertheless, nobody knows what the males achieve by being so noisy. It does allow them to know each other's position and perhaps to divide up the forest between them. Also, there seems little doubt that all the local orangs can recognize the individual voices of the male callers, but it certainly does not serve to drive all other males from the caller's patch. Indeed, it may allow them to remain, safe in the knowledge that the caller advertises his position often enough for them to avoid him. It may be that receptive females are attracted to the callers, but females sometimes move away from the sound or hide from adult males.

For much of their lives, each sex shuns the other, and even when they do meet, they seldom consummate their relationship right away. Subadult males tend to show very much more enthusiasm for sex than is the case with adults, something that is almost perfectly contrasted with the females' lack of enthusiasm for them. They often follow females around and may attempt to put their arms around them or to lick their genital areas. The females may resist or move away, but subtlety is seldom a male's strongpoint. With four hands and superior strength, he has the ideal build for a rapist and many an adolescent or adult female is taken in this way. She may cry and scream, or utter low, guttural noises that she gives at no other time, but the male is often successful. Sometimes, the female has an accompanying youngster who attacks the male in his or her mother's defence, but this is totally ineffectual. The best that most rape victims can hope for is to struggle so effectively that the male cannot effect penetration and ends up rubbing against the outside of her body.

Adult males, as well as subadults, sometimes rape females. From a reproductive point of view, it might be thought that the males are increasing their chances of having offspring, especially if they are subadults with little chance of seducing a female in a more gentlemanly fashion. However, the females almost invariably fail to become pregnant as the result of a rape. Amazingly, this is not because the male fails to ejaculate or does so in the wrong place, although both often happen; it is something that the females control with their bodies.

It seems that most, if not all, pregnancies are the result of willing matings by females and that these generally take place during consortships of several days' or weeks' duration. During these relationships, the couple usually precede sexual intercourse with imaginative foreplay on the part of whichever partner is the initiator at that moment. Adolescent females are often very much keener than their older mates and may hang, legs apart, in front of the male's face. Such a young female might also hug her lover and apply her hand or her mouth to his genitals in the apparent hope of raising his interest. Oral sex seems to be particularly popular among orang couples.

Intercourse takes place either face to face, or with the female facing away from the male. Often they both hang beneath a branch, sometimes for a leisurely 20 minutes. But the most reproductively successful matings seem to be those in which the male practically lies on his back in the trees. The female then straddles him and thrusts up and down until they are finished. For some reason (possibly a lack of female cooperation), subadult males do not mate like this.

It seems quite likely that the reason why females seldom get pregnant from rapes but readily do so from mutually motivated sex is that they only have orgasms during the latter. It has been speculated that contractions of the female's uterus during orgasm suck the semen in the direction of the egg that is waiting to be fertilized.

As the males grow older, so they seem to lose interest in sex long before they become really ancient. Probably in their early twenties, when they are still strong and intimidating specimens, they seem to give it up altogether, even hurrying away from amorous females. Why this should happen is not clear, but it presumably has something to do with the great weight and decreased mobility of these animals. Big males are the only orangs that spend a great deal of

Mother and infant orang-utans spend most of their time with only each other for company, in contrast with most other primate species which are more social.

time on the ground, apparently because it is much easier not to climb about in trees when you weigh close to 100 kg (220 lb).

The question that the evolutionist must ask is why do these old males go on living? Are they useful to their offspring in some way? There is no easy answer, but it could be that an old male somehow conveys information – perhaps about food at a critical time – to his maturing offspring. Alternatively, he may be tolerant of his sons in some way and protect them by his presence from the arrival of another big, resident male until one of them is big enough to take over. The real reason why these questions cannot be answered is that, for all that is known about orangs, they are still among the least understood of the great apes in almost every aspect of their social lives. Only the pygmy chimpanzee has more secrets.

Nevertheless, they have been well-known in captivity for many years due to the ease with which babies could be obtained by shooting their mothers. Those people who have been prepared to collect them in this way seldom bothered about the babies that died as a result of their inevitable fall from the trees, or from shock, or from the maltreatment that usually followed. And sadly, the zoos that bought them were seldom concerned about their origins. Nowadays, zoos tend to be rather embarrassed about this and the governments of Malaysia and Indonesia do more to protect their apes, so the orangs have a hope of surviving as long as they have a forest home. Once that has gone, they really are doomed as captive-bred individuals almost never produce a subsequent generation.

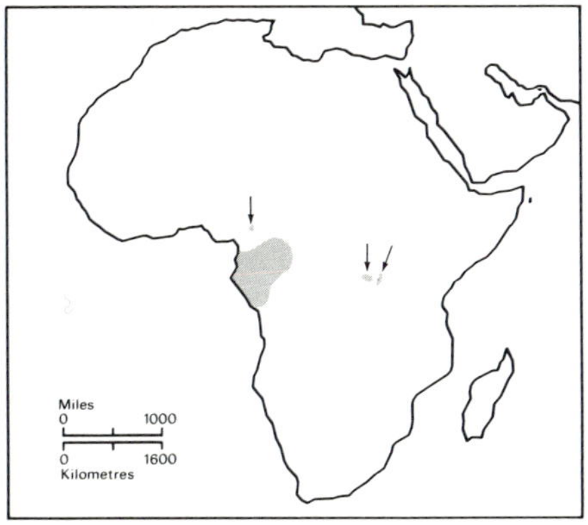

The gorilla GENUS *Gorilla*

On a cold, wet, foggy morning, on a high mountain slope in Central Africa, a channel of crushed vegetation permits easy passage through the grass and shrubs. Occasionally, the channel forms a tunnel through places where vigorous young trees and bushes grow particularly densely. In other places, it winds over logs or under strangely gnarled old trees with branches festooned with sodden lichen. From all around, occasional deep grunts penetrate the gentle sound of the wind in the tall grass, and sharp cracks and crunching sounds form a constant background accompaniment. Every now and then, a hairy black arm reaches into view and the hand on the end of it grasps a piece of vegetation and pulls it back into concealment.

The owner of the hand is a mountain gorilla, one of a group that is peacefully foraging for its breakfast, totally oblivious of the mixed esteem, awe and horror with which the human primate views the biggest species in the order.

Three races of these powerful apes inhabit the forests of West and Central Africa, from Cameroon to Uganda and Rwanda. The

Adult male gorillas are known as 'silverbacks' from their pelage colouration. This mountain gorilla is in the prime of life.

smallest race (if that is an appropriate description) is the western lowland gorilla which inhabits parts of the great forest block between western Cameroon and the Congo (Zaire) river. Adult males average about 168 cm (5 ft $5\frac{1}{2}$ in) tall. As in all three races, the females are little more than half the size of the males. Western lowland gorillas have relatively broad faces with smaller jaws and teeth than their eastern relatives. Their fur is short and generally blackish-grey, although distinct brown shades are also common. Adult males have a light 'saddle' of nearly white fur across their backs, a feature common to all races, from which the popular term 'silverback' for a fully mature male derives.

There are two eastern races of gorilla which seem to intergrade into one another to a certain extent, although most of the populations are now found in isolated pockets of suitable habitat. Typical mountain gorillas are found in the Virunga range of extinct volcanoes that straddles the borders of Zaire, Rwanda and Uganda. Contrary to general belief, this is not the biggest race, although its appearance is somewhat enlarged by long, silky, black hair. Males weigh about 155 kg (343 lb) and are said to be about 172 cm (5 ft 7 in) tall. There is, however, a possibility that they are really rather stockier than these measurements would indicate, being perhaps this heavy but shorter in stature. Like the western lowland gorillas, they have broad faces, but their jaws and teeth are particularly massive.

The largest gorillas in the world – and therefore the largest primates – are those of the eastern lowland race. Silverbacks average about 165 kg (363 lb) in weight and 175 cm (5 ft 8 in) in height. They may be more easily distinguished by having rather long faces and short, black fur. This race is found in parts of the lowland forest of eastern Zaire and in some of the adjacent highland areas near the international border.

Animals often fail to conform to nice, neat biological categories, so it is not surprising to find that the gorillas of one mountainous locality to the west of Lake Kivu in Zaire are a kind of mountain gorilla yet slightly like the eastern lowland race.

Just as among people, exceptional gorillas grow much bigger than the average, and the largest wild one ever weighed is said to have been one of the mixed race from near Lake Kivu that turned the scales at 210 kg (462 lb). Many zoo animals can beat this easily, but that is because they become disgustingly fat if kept in cell-like conditions with nothing to do. The record is an appalling 350 kg (772 lb)! A healthy male gorilla should have a hint of a potbelly, but the overall impression should be one of well-toned muscle, not flab. His crowning glory is a conical mass of bony ridges and musculature on top of his head.

As gorillas are found from sea level to 4000 m (13,000 ft) above that, it is not surprising that they occupy a number of very different types of vegetation, from hot, humid, equatorial rain forest to temperate, herbaceous hillsides and lichen-covered, small-statured forests of misshapen alpine trees. Mountain gorillas often have to face freezing temperatures at night, and spend almost their entire lives in damp conditions, mainly between 2800 and 3400 m (about 9200 to 11,000 ft) above sea level.

The cold conditions in the Virungas have led to the novel theory that the reason why the gorillas from this area regularly defecate in their nests and sleep on their dung is to provide a little extra insulation. With the exception of dependent babies, gorillas usually build nests for themselves in which to sleep at night, but most of them are rather more sanitary than those of the Virungas. The idea is that because the dung is dry and rather fibrous, it makes a good insulator, but it is only attractive to sleep on in conditions of extreme cold.

Unlike the other great apes, gorillas do not always nest in the trees, being able to make quite springy beds out of the ground layer of vegetation, both for sleeping at night and sometimes for a noon siesta. The height at which they nest seems to depend mainly on the type of material available, and probably on individual preferences.

Foraging heights also vary greatly between the gorillas of different localities. In the Virungas, for example, copious amounts of ground-level vegetation mean that the gorillas there can be almost entirely terrestrial feeders. They do climb into the trees to collect certain items, such as particular fruits and flowers, but they are distinctly cautious in the way they go about it. In some other areas, where more trees and vines are available, considerable climbing skills are shown and gorillas frequently feed way up in the trees, sometimes dangling casually beneath branches. Even the biggest silverbacks can be seen climbing some of the time, although it takes a strong tree to support weights in excess of 150 kg (330 lb). Not surprisingly, the females and younger animals tend to be the more arboreal members of the groups.

It follows that there are differences in the diets of smaller and larger animals, but all can be summed up as being almost entirely vegetarian and including an extraordinarily high proportion of leaves, shoots and stems for a non-colobine primate. In some areas, where the available plant foods are extremely diverse, no animal matter is eaten; but where the vegetarian diet is simpler, grubs may be included in the diet, perhaps to offset particular vitamin deficiencies.

Gorilla groups have home ranges of anything from 4 to more than 30 km² (1.5 to 11.5 square miles). The area that is most used by each group tends to be fairly exclusive, but there is no systematic territorial defence such as occurs among chimpanzees, for example.

Foliage makes up a large part of the diet of this female western lowland gorilla.

The numbers of animals in the groups are quite small, varying between 2 and about 20, with a tendency towards smaller groups among the western lowland race and larger ones in the mountains.

The species is one of those rare ones in which the focus of the group, the unifying force, is the leading male. When slowly on the move and foraging, the group tends to spread out, but when relaxed, the leading silverback is the centre of attraction for the females and young. He may sometimes find it necessary to attract their attention with noisy displays of chest-beating, ground-beating, tearing up vegetation and short, sudden, sideways rushes – and he might even add a little mild female gorilla-beating for good measure – but it is almost invariably they who move to sit by him, or to follow him, not vice versa. His attention-getting displays are most likely to be given when it is time to move on after a rest period and seem to be best interpreted as 'Everybody wake up and get ready to follow me!' When something disturbs the group and they become fearful, the females and young will need no reminder to cluster around their leader.

The life expectancy of a healthy wild gorilla is unknown, but various guesses place it as high as 60 years; and even if that seems unlikely, it must be fairly long, since males do not become fully mature until they get their silver backs at the age of about 11 or 13. They become adult in most other respects at the age of about eight, after which they go through a 'blackback' period. During this time, they are not socially mature, they do not lead groups and they spend increasing amounts of their lives apart from the other members of the group. However, they remain playful in nature and some individuals manage to maintain a fairly close relationship with the leading silverback. He in turn is likely to be the blackback's father.

Upon reaching silverback status, a young adult male does not necessarily strike out on his own. He may do so eventually, but he will be more likely to remain at least loosely attached to his parental group. Some young silverbacks wander off for days or weeks, returning later; others gradually encroach on the leading role. Gorillas are highly intelligent creatures and it is not surprising to find individual differences and preferences at work in almost every facet of their lives. If a young silverback does stay in his original group, the chances are that he will be able to sire some offspring, particularly with the younger adult females, and he may eventually take over as Number One.

Females mature quicker than males, as is so often the case, becoming adult at the age of about eight. This is when they have their first sexual swelling, a small phenomena in this species and something that marks the period of peak sexual receptivity, as in many primates. It presumably marks the period of ovulation. Some females mate at other times as well, including when they are pregnant; and, once in a while, two females will 'mate' homosexually with each other.

That notwithstanding, the females seldom seem to be particularly attracted to each other. They typically leave their parental groups as young adults, although sometimes they produce an infant before they go. They then make almost immediate transfers either to solitary silverbacks or to other, existing groups. In this way, new social units are formed and old ones change in size.

Young mountain gorillas often ride on their mothers' backs.

The relationships between different groups are extremely variable. At their most peaceable, two that know each other well might mingle and even pass a night together, later splitting up again into their original units. On the other hand, some groups – or, more likely, some silverbacks – just do not seem to be able to stand the sight of each other. When they meet, noisy displays, with the usual chest-beating, rushing about and so on, take place between the big males, with the additional exchange of loud hoots. Such encounters can thoroughly disturb one or both groups, often resulting in one moving away for a time from its usual haunts.

Fierce fights can sometimes erupt between opposing silverbacks and weak leaders may be severely challenged. Lone outsiders can sometimes be a particular threat both in terms of their potential for attracting away females and because there is good reason to believe that any unrelated silverback might attack and kill the infant of a potentially attractive female. Infanticide is always hard to document, but, just as in the langurs, it seems that the mother of a victim is quite likely to respond to her loss by mating with the killer. No moral judgement is appropriate here for the mother is small, unable to defend her baby against a massive male, and she must get on with the perpetual business of breeding. The male, for his part, can give no quarter in his competition with others and stands to gain by the mother's accelerated receptivity. In his turn, he will become a devoted father.

It is far easier to pass a judgement on the humans who threaten the very existence of all gorillas. Poverty may aggravate both agricultural encroachment into their last remaining forests and poaching in national parks, but the damage that is done is often catastrophic all the same. A wire snare left for an antelope may catch the limb of a passing ape. It will not hold the animal, but it will cling on to it and cut into the flesh. Pulling at it only draws it tighter, and the result is usually gangrene at first, then death. Most inexcusable of all, some perverted people from the wealthy, industrial Western countries have encouraged the killing of gorillas by offering large sums of money (fortunes to the killers) for such macabre trophies as their heads and hands. Beside such calculated brutality, the occasional outburst of fatal violence in the generally peaceful gorilla society pales into insignificance.

The chimpanzees GENUS *Pan*

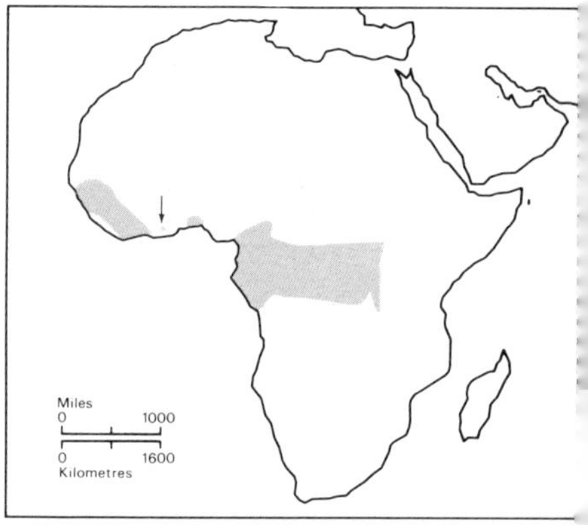

A big, black ape walks slowly on all fours up a large, inclined branch and pauses about 15 m (50 ft) above the forest floor. He reaches out to another, higher branch with one hand and casually swings his muscular body onto a convenient fork. Once there, he begins, with methodical deliberation, to bend in towards himself all the leafy branches and twigs within his reach. He holds each one with his feet to prevent it from springing back, and then sits on them to restrain them more firmly. The growing structure looks as though it will burst apart the second that he releases the tension, but after about two minutes of building, he seems content and lies down on his side.

He has built himself a day nest, a springy mattress in the trees that will keep him comfortable while he dozes for an hour or two. Nearby, other chimpanzees are doing the same thing, while a few merely sit on big branches, heads on chests and eyes closed. Two young infants, with the little white tail tufts that signify their youth, play a hesitant game of tag beside their mothers; and elsewhere a half-grown male stares down at some terrestrial animal that has captured his curiosity. It is seista time for a group of Ugandan chimpanzees; but the scene, or one like it, could be occurring in any one of thousands of locations from Senegal to Tanzania.

The presence of chimpanzees in an area can always be detected by the nests that they build in the trees for sleeping by day or by night. Apart from those youngsters that are small enough to share with their mothers, each individual builds a new nest each time he or she wants to use one, so a forest that contains chimps also contains hundreds of new and old nests in various stages of decay.

Three very similar races occupy the forests and some of the denser woodlands across a broad belt of Africa. They may be distinguished from one another mainly on the basis of their slightly differing complexions, but it seems to be unlikely that the divisions correspond to any behavioural or ecological differences. The westernmost race has a dark mask surrounding its eyes on a light face, the central race has freckles and the eastern chimps have rather pinkish faces when young; but all become darker as they get older, making it difficult to tell them apart.

They live in complex groupings that are like those of spider monkeys in some ways; groups are hard to identify because their members travel around individually, in pairs, or in almost any conceivable combination of the members. Because males and females seem to have rather different ways of 'belonging' to the group, it is probably better to refer to the social units simply as 'communities', which may have almost any number of members, up to about 80.

Forest-dwelling chimps are not well known, but in the woodlands beside Lake Tanganyika, in the Mahale Mountains and in the famous Gombe Stream National Park where Jane Goodall has studied them, the community is based on a clique of males that live in a territory which they defend with great hostility against others. Parties of males, especially young adults, patrol the boundaries, apparently looking for members of the adjacent community to challenge. If they meet an opposing patrol, one group or the other may discreetly make off, or the two parties may leap about, break branches and scream at each other in a dramatically noisy display. In a very few, isolated instances, such boundary patrols have been known to attack lone males of a neighbouring community with savage ferocity. The victims have been sat on, stamped on, hit, bitten and pulled about, being left finally in a state of deep shock and severely wounded. Some such victims are believed to have died; but before these thuggish tactics are assumed to be normal for chimpanzees, it must be remembered that the apes concerned had been in close contact with humans for years and had had at least some of their behaviour altered by regular hand-outs of food. Among many primates, it is easy to start a quarrel or a fight by giving them a large quantity of food at one place, and who knows what effect this might have upon such complex and intelligent beings as chimpanzees?

The males are clearly permanent members of their communities, growing up in them and attached as much to each other as to the land, but the position of the females is rather more difficult to define. They certainly behave on a day-to-day basis and have a variety of social relationships with each other and with the males just as

LEFT Black faces and often a rather balding appearance are typical of adult chimpanzees. This female comes from Uganda and is a member of the eastern race.

CENTRE The dark area in the middle of the picture is a nest that a chimpanzee has built for a midday rest. The occupant can just ben seen.

BOTTOM Chimpanzees are large apes and they are often very noisy. But they can blend into their surroundings and be quietly inconspicuous when they want to be.

would be expected of community members. They even tag along with the males on boundary patrols, being especially likely to do so if they are at a sexually receptive stage – something that they signal with a large, pink swelling, like many monkey species. However, they also move between communities, seeming to be most likely to do so before they have their first baby.

Female chimpanzees have regular periods of sexual receptivity for months before they conceive their first infant, and it may be that this is an important time for them to try out different communities before they settle down. It is unlikely that they would be received with any great joy by the resident females they encounter, but the males are likely to be happy enough at the arrangement to be both welcoming and protective.

When fully adult, females are rather less gregarious than males, concerning themselves mainly with their offspring, with whom they maintain close relationships for years. (It takes about eight years for a chimpanzee to reach puberty.) An exception to this occurs during their mating periods, but in the main, they travel less far and socialize less than the males. It even seems that once a female has settled down as an adult, she does not move if the neighbouring male clique pushes its boundary into her males' territory, past the area that she regularly frequents. In such a case, she will probably accept the change of males imposed upon her. Thus, it seems that a male clique can measure at least part of its success in terms of the number of females that live within its boundaries; and, like individual males of several prosimian species, an increase in territory size is likely to bring an increase in the number of available females.

Within the male clique, different males have different amounts of success with the females, but really fierce competition is probably kept in check by the need for them to present a united front to their neighbours. Quarrels within the family can only go so far.

At 40 to 50 kg (about 88 to 110 lb), male chimpanzees are about 12 per cent bigger than the females, and there are differences in their behaviour that go beyond their social lives and reproductive functions. Although chimpanzees depend on fruit as the mainstay of their diet, they also eat regular amounts of animal prey in the form of mammals, birds and insects. These are eaten frequently enough to be important sources of protein, but the females eat more of the insects and the males more of the meat.

Termites form a regular item on the insect menu. These social, antlike insects live in large mounds of their own construction and are collected by means of a form of fishing. A chimpanzee takes a long blade of grass, a thin piece of vine, a twig or something similar and modifies it appropriately. It may be shortened, or stripped of leaves or branches, or whatever is necessary. Having made the tool, it is inserted into a hole which the ape (or another) has previously dug into the wall of the mound. The termites oblige by locking onto the alien object with their mandibles. It is then a simple matter for the chimp to pull up the tool carefully, so as not to knock off the termites, and pluck them off with the lips and eat them.

Because termites remain living in individual mounds for a very long time, chimpanzees can fish for them at the same place year after year. They thus have a regular supply of insect food; and although termite fishing does not take up much of the day, its reliability as a source makes it an important dietary activity. Females termite-fish more frequently and in bouts of longer duration than the males. They might spend over 4 per cent of their waking time in the activity, whereas males average less than a third of that.

Migratory driver ants are also collected by dipping with a long implement, but these are faster, more aggressive creatures that swarm up the tool and the fishing ape has to move fast to sweep them off with one hand and stuff them into his or her mouth. Because their nests are temporary bivouacs that are only found by chance, driver ants do not form such an important item in the diet as termites, but, here again, females seem to take more of them than do males. The same also seems to be true of weaver ants, which are among the most popular of insect foods. They make arboreal nests out of leaves that are woven together with silk, and it is easy enough for a chimp to find a nest, crush it and eat the occupants.

Meat is obtained either by opportunistic collection, as is the case with eggs and nestlings, or by hunting it, as is the case with most mammalian prey. In addition, it is sometimes stolen from baboons. More often, young baboons themselves form the prey, as do redtails, blue monkeys, red colobus monkeys, small antelopes and bush pigs. Even a bat has been known to fall victim.

Given the opportunity a chimp will grab a prey animal that is careless enough to put itself in danger, but many of them are caught after careful stalking and chasing. For example, small monkeys are isolated in clumps of trees while members of the hunting party station themselves so as to cut off all escape routes. One chimp then climbs after the monkey and with luck (but not for the monkey) it is caught.

Almost without exception, it is the adult males that initiate hunts and capture the prey. As a result, they eat most of the meat but it does tend to be shared, particularly with animals that beg for it, and females in the peak of their sexual condition do rather well as recipients.

Some anthropologists have repeatedly suggested that share-outs at the successful conclusion of hunting forays were the origin of human cooperation and that hunting was therefore the basis of our cultural development. However, as chimpanzees frequently share plant food, such as large fruits, between mothers and their offspring, and food sharing is found among vegetarian monkeys and lemurs, the hunting hypothesis seems rather weak. What the chimpanzee example shows is the beginnings of an ecological division between the sexes in which the females are the collectors and the males the hunters. Furthermore, the females seem to have taken the lead in tool use.

Other evidence of the talents of this remarkable ape come from the studies that have been made of it in captivity. Young chimps have long been known to have the insight to solve such problems as the need to fit sticks together to knock down an otherwise unreachable food reward, but the species' remarkable language abilities are perhaps less well known. Attempts to get chimpanzees to speak have never produced a vocabulary of more than four rather breathless words, which would hardly do credit to a parrot. However, attempts to teach visible languages (rather than auditory ones) have been much more successful.

An important beginning was made in the late 1960s when a six-year-old chimp was taught to arrange and to read plastic symbols in a meaningful order on a board. Each symbol corresponded to a word and the ape showed the ability to learn a simple code, which she could use for such exercises as asking for a piece of apple.

Since then, two research projects have produced chimpanzees that can use languages possessing numerous 'words' and grammatical rules. In both cases, the chimpanzees can understand original combinations of 'words', as long as they follow the grammatical rules, and they can construct completely new strings of 'words' that make perfect sense. This is equivalent to understanding or saying a totally original sentence, using familiar words and the standard rules of grammar, in English, French, Malay or any other language.

In the more precise of the two projects, young chimpanzees at the Yerkes Regional Primate Research Center in the United States communicate with a computer console in a specially invented language called 'Yerkish'. The 'words' are symbols on computer keys which can be punched in the equivalent of speaking. By means of more than one console and a visual display of what is punched in, the chimpanzees can communicate with each other and with the experimenters about such important matters as food, drink, music and being tickled. Requests have to include the word 'please' and sentences must end with full stops.

The disadvantage of talking to a computer is that they do not yet show the imagination of those in *Star Wars*, so the chimps at the Yerkes Center are strictly limited in what they can invent. No such strictures apply to the chimps that have learned American

The half-grown male Ugandan chimpanzee still has a pale face, but he already moves quite independently of his mother.

Sign Language in a project that began at the University of Nevada and is being continued at the Institute of Primate Studies in Oklahoma. American Sign Language is a gestural language used by the deaf: to 'speak' you make specific gestures for specific meanings.

Washoe, the first chimp to learn American Sign Language, acquired a vocabulary of more than 130 gestures in four years of learning. Since then other apes have done equally well or better, although there is great individual variation in their language capacities. Some of them can now chat merrily to people or other chimps in the sign language. When they do not know gestures, they sometimes invent new ones. Washoe, for example, invented a sign for a bib by making the appropriate outline on her chest. Alternatively, original combinations of signs can be used to overcome the lack of a specific one; the creative Washoe referred to nuts as 'rock berries'.

All this is way beyond the imitations of a parrot, but the prize goes to Washoe, once again, for inventing swearing. She had a sign for 'dirty' which she commonly applied to faeces. Following an altercation with a caged rhesus monkey (not in sign language!), she applied it to the monkey, since when she has freely applied it to people who earn her displeasure. Most people would have thought this inconceivable only a few years ago. Yet, since these chimpanzees have been taught to communicate in this way, a few gorillas have been persuaded to take up American Sign Language and it is hard to see where the whole process will stop.

Remarkable and familiar as the chimpanzee is, it is not the only member of its genus. In the remote forests of central Zaire, there lives a completely separate species, the pygmy chimpanzee or bonobo. This species is very like its better-known cousin, but rather smaller in stature and lighter in build. Although it does not weigh much more than half as much as the latter species, the bonobo can easily be mistaken for an undersized ordinary chimpanzee, which is why it was not discovered until 1925 when an observant anatomist became intrigued by some skins and skeletons he found in a Belgian museum. Since then, it has been realized that the lighter build of the bonobo is an adaptation that enables it to spend more time in the treetops than the ordinary chimpanzee and to be more agile when it is there. Ordinary chimps can hurl themselves about among the branches when suitably motivated, but most of the time they are rather careful and precise in their movements, in contrast to the casual grace of the bonobo. On the other hand, the larger species is better at moving long distances on the ground, something that is very necessary wherever it occurs in woodlands rather than in forests.

Ordinary chimpanzees and bonobos are separated by the huge Zaire river, with the smaller species living to the south of it. The political and practical difficulties of going to its habitat to study it mean that it is the least known of all the great apes and, sadly, it seems to be declining rapidly in the wild. It has recently disappeared from some forests in which it used to live and only a few thousand individuals remain, scattered in isolated patches where neither land clearance not hunting for the pot have yet taken their toll. Unfortunately, bonobos do not occur in even one single national park, so unless something is done quickly, we never will know much about them.

11 The Hominids: upright revolutionaries

One of the reasons why our species is so interested in the other primates is that by looking at them we can obtain some idea of what our ancestors must have been like a few million years ago. Although we are not descended from any modern-type monkey or ape, our lineage does appear to have gone through stages in which we were a medium-sized, reasonably intelligent creature with good binocular vision, hands that were good at manipulation, and the ability to climb trees. Detailed speculation on how our ancestors might have lived would require another book, but it is not difficult to see that they must have been forced to adapt to their environment in ways that did not differ all that radically from the other primates.

Anthropologists might be less interested in comparisons with monkeys and apes if the direct evidence about the nature of early hominids (members of our family) were a little better than it is. Unfortunately, it all has to be dug up, quite literally, and with millions of square kilometres of land in the world in which to dig, a few scattered fossils are not easy to find. Of course, certain areas, such as places that have consistently been extremely cold for millions of years, can largely be ignored; but that still leaves a vast area in which to search.

Because of the difficulties, it is not possible to be certain about the course of hominid evolution, but enough evidence has been drawn together for what follows to be a plausible summary.

The most recent common ancestor of hominids and apes that has been discovered in fossil form was a small, agile, tree-dwelling ape of vegetarian disposition. It belonged to the genus *Aegyptopithecus*. This creature lived about 28 million years ago in what were then the lush forests of North Africa and retained certain 'pre-ape' characteristics, probably including a tail.

The ancestral lines of the apes and humans probably diverged less than 25 million years ago at a time when several species of the early ape genus *Dryopithecus* inhabited Africa, Europe and Asia. Our last common ancestor may have been one of the first species of *Dryopithecus* to have evolved. It was almost certainly small and arboreal, like *Aegyptopithecus*, and may have lived a similar sort of life to that of the modern pygmy chimpanzee, foraging all year round on whatever fruits, flowers and young, tender leaves were available.

It should be apparent by now that our affinities are clearly with the Old World primates and that our spread into the Americas is something that happened after mankind evolved its separate identity. By about 14 million years ago, the progenitors of today's great apes had reached the peak of their geographical distribution and were spread from China in the east to Spain in the west, and from Germany in the north to some as yet undetermined point in Africa south of the Equator.

At the same time, our ancestors were probably doing quite well by feeding both on the ground and in the trees in the more open areas, either within the forest itself or at its edge where it gave way to woodlands and savannahs. These more open habitats were spreading in Africa due to a general drying out of the climate, a phenomenon that probably gave the crucial push to hominid evolution by favouring the spread of the genus *Ramapithecus*. Tantalizingly few remains of this genus have been found, but they probably based their success on being better at travelling across open country and at eating the roots, seeds, nuts and berries they found there.

An important indicator of their affinity to modern man is that *Ramapithecus* species did not have the big, daggerlike canine teeth of the great apes and most monkeys. They had teeth that formed a fairly even row, much like ours. Current speculation is that this was an adaptation to the rotary jaw action that is necessary for chewing up the many tough roots and seeds that are there for the taking on the grasslands of the drier tropics. Our own teeth are certainly very well adapted for chewing since we have broad cheek teeth for grinding and crushing, and no projecting canines to get in the way. Ethiopia's grassland-dwelling geladas provide us with something of a modern parallel in as much as their cheek teeth also have broad surfaces, although they have retained their big canines.

Geladas also have one of the best precision grips of any non-human primate, and by looking at the way in which they pick up the small objects on which they feed, we can infer that *Ramapithecus* must have been under similar evolutionary pressure to be clever with his or her hands. We really do not know how *Ramapithecus* lived or even how well he walked upright, but somehow the spread of the savannah-woodlands at the expense of the dense forests favoured primates that could move easily on the ground, pick up small objects and chew them up. Geladas, baboons and hominids have all made this transition, but *Ramapithecus* went furthest and lost his big canine teeth long before his descendants hit upon the idea of using sharp stone tools as substitutes.

Ramapithecus individuals were not much over 1 m tall (about 3.5 ft) but they probably did not suffer greatly from the attentions of predators as a result of the absence of any fearsome dental weaponry. Today's well-armed baboons may fight with predators on occasion but they also provide ample illustration of the adaptive value of running away at the right moment. *Ramapithecus* was probably very good at dashing for the nearest tree. Given his ability to hold objects and to stand up straight at least some of the time, he was probably also quite good at bluffing medium-sized predators by brandishing and throwing large sticks. Modern chimpanzees are perfectly capable of giving very intimidating displays along these lines.

Since we know that various monkey species can modify their behaviour quite radically over short periods of time in response to changing conditions, it seems only reasonable to suppose that the behaviour of *Ramapithecus* was similarly flexible. It would be surprising if our early ancestors were not at least as good as macaques at developing new ideas by invention and copying; and Kenya's olive baboons provide an example of how quickly an open-country primate can incorporate cooperative hunting into its repertoire. It would be equally surprising if *Ramapithecus* did not make some use of tools, at least to the extent of fashioning sticks to extend his reach in the manner of chimpanzees. Thus, although we do not know exactly how *Ramapithecus* lived, his primate heritage clearly gave plenty of scope for intelligent adaptation to the local conditions. As a result, three or more species of this genus spread out over much of the Old World, appearing in Asia no less than 11 million years ago, and in Europe about one million years later.

What happened next is rather confusing; and with important new fossils being dug up at the recent rate of about one a year, conflicting interpretations and re-interpretations abound. What is certain is that *Ramapithecus* was succeeded in Africa by more than one different lineage of bipedal hominids. Unfortunately, the relevant fossil record is effectively non-existent for the period between about 12 and 4 million years ago. On this side of the gap, some late representatives of *Ramapithecus* are still around but they have been joined by a variety of creatures of the more man-like genus *Australopithecus*. These were hominids that were a little larger than *Ramapithecus* and had bigger brains (probably between 400 and 600 cc, in comparison with modern man's range of 1000 to 2000 cc).

There were at least two lineages of *Australopithecus*: a stocky, 'robust' line that was about 1.5 m (roughly 5 ft) tall and a slightly shorter and slimmer, 'gracile' line. Because of the physical differences between them, they must have differed in their ecological adaptations, but both did well for a time and appear to have been widespread in Africa between about four and one million years before the present.

The robust line may have lasted until considerably more recently than one million years ago, but it eventually died out without leaving any descendants. The gracile line appears to have gone sooner, but many anthropologists believe that it gave rise to the genus *Homo*. An opposing school of thought points to nearly four-million-year-old East African remains of a highly evolved creature called *Homo habilis* and argues that it must have evolved in parallel with *Australopithecus*, making the latter genus a blind alley. It is safest to say that we *might* have had some *Australopithecus* forebears and that, if so, they *might* have been early gracile types but they certainly were not robust ones.

What we do know is that the earliest members of our genus arose in Africa and shared their environment with at least three other hominids: one late representative of *Ramapithecus* and two *Australopithecus* species. Seeing how many closely related monkeys, apes and other mammals often live together, it is not really surprising that there should have been more than one hominid species at one time. All the same, it is a little eery to think of a world in which we might meet creatures that are so nearly human without quite being so.

The oldest yet discovered member of our genus, *Homo habilis*, was about 1.25 to 1.5 m tall (roughly between 4 and 5 ft) and had a brain of about 800 cc. That is more than half the modern human's average of 1400 cc. Very little is known about *Homo habilis*' life style, but they seem to have had limbs that were very like ours. Thus, they probably walked as easily as we do and gripped objects almost as well. On both counts, they probably scored heavily over *Australopithecus*. We also do not know how long they lasted, but by just over $1\frac{1}{2}$ million years ago, the next *Homo* species, *Homo erectus*, had arisen, also in Africa. They coexisted for half a million years or more with the *Australopithecus* species but eventually became much more successful.

The main factors that differentiated primates of the genus *Homo* from those of *Australopithecus*, and which contributed so much to the success of the former, are their greater systematic production and use of stone tools, the establishment of home bases in the form of temporary camps to which individuals returned at night, and the regular sharing of food. These three factors, especially the last, were at the heart of our complex economic and social evolution. And the more it developed, the more humans had to gain from sophisticated cooperation with each other and, thus, from their psychological development.

Africa has aptly been called the cradle of mankind, but about a million years ago, *Homo erectus* climbed out of the cradle and spread throughout Europe and Asia. He was not so different in stature from modern man, but if he were to appear today, he would be readily distinguishable by his big, protruding jaw, his prominent, bony brow ridges and his backswept forehead. His brain size would be in the range of 775 to 1300 cc, overlapping in volume (but not necessarily in intelligence!) with our own species.

No species evolves from its immediate ancestor in a single, breathtaking jump but *Homo sapiens* can be said to have arrived about half a million years ago. Since then our species has spread out into the widest distribution of any vertebrate and proliferated into a variety of races. The well-known Neanderthal man of Europe and the Middle East, for example, was a stocky, cold-adapted race which flourished during the last ice age before disappearing about 30,000 years ago. The Neanderthals probably succumbed to a mixture of being killed, pushed out of the best land and assimilated by modern humans.

This has been a thoroughly sketchy introduction to the origins of the last primate genus and interested readers are urged to consult some of the authoritative works listed in the bibliography. The intention here has been to show something of the evolutionary pathway from non-human to human primate, because so many of the latter find it easy to forget their ancient heritage when considering their biology, their psychology, their sociology, their economy and just about any other aspect of their place in the modern world.

The human GENUS *Homo*

Although there is only one human species alive in the world today, its sheer diversity defies the simple sort of encapsulation that has served to introduce the other genera. The description of its ecology and behaviour requires disciplines such as anthropology, agronomy, economics, psychology, sociology and so on, all of which are therefore branches of human biology.

With a world population of about four billion, *Homo* is not only the most numerous of primates, its numbers exceed those of all the other genera combined. Yet all individuals are commonly referred to as belonging to the single subspecies *Homo sapiens sapiens*. (The second '*sapiens*' is the name of the sub-species.) On the face of it this is surprising, since minor physical differences, such as in pelage colour, are frequently used to justify the naming of new subspecies of non-human primates, and there is no obvious difference between the two categories 'race' and 'subspecies'. The essential criteria for recognizing them are that they should be identifiable (usually as museum specimens) and that they should come from different places. When two formerly separated subspecies of non-human primates come into extensive contact with each other, they may be expected to interbreed and the difference between them to disappear. If that does not happen, then they should be regarded as belonging to different species.

It could be argued that many of the different races of man are consistently different enough to be regarded as being of separate subspecies, and this would be in line with the way in which we have treated other primates. However, even if this is the case, there is no doubt that the ease of modern long-distance travel is promoting more and more opportunities for humans to demonstrate their membership of a single species by interbreeding freely with one another. The long-term result should be the disappearance of discrete races.

7. PROSPECTS FOR THE PRIMATES

There are 182 species of non-human primates alive in the world today. Most of them are monkeys; there are 128 monkey species, 51 of which occur in the New World and 77 in the Old. The prosimians are the next biggest group with 41 members, and finally there are 13 different ape species. Twenty-eight of the world's primates are found on Madagascar and the nearby islands, and the remainder are almost evenly divided between Africa (51 species), Asia (53) and the Americas (51). One species, the hamadryas baboon, is counted twice because it appears in both Africa and Asia.

The vast majority of primates occur in the moist forests of the tropics where they are among the most conspicuous of the mammals. Any healthy tropical forest in Africa, Asia or America has its population of primates, which even a casual traveller may occasionally glimpse among the trees. Only a relatively few species have adapted to life in more open habitats, or in colder places. However, those species that have moved out of the forests, including the green monkeys and some baboons and macaques, are among the most numerous and therefore the most successful of monkeys.

Consideration of all the genera has shown that primates universally lead quite complex social lives, although a few of them do not actually spend a great amount of time in contact with other members of their own species. Social systems in which individuals or mothers and infants form the basic unit are found among the prosimians and the orang-utans. In all these cases, the animals concerned are still 'social' in so far as they have complex and generally competitive or sexual relationships with their neighbours. It is likely that in all species, these primates normally know their neighbours as individuals and are aware of their strengths, weaknesses and sexual conditions. In addition, relationships between parents and offspring may well involve assistance for the younger generation well into adulthood.

All of the other primates are group-living in units that vary from monogamous families to highly complex, variable communities. Monogamy occurs among some of the prosimians of Madagascar, probably among the tarsiers, among several South American

SUBORDER
STREPSIRHINI

	GENERA	SPECIES
Superfamily **LEMUROIDEA**		
Family **Cheirogaleidae**		
Subfamily Cheirogaleinae	*Microcebus*	*murinus* grey mouse lemur *rufus* rufous mouse lemur *coquereli* Coquerel's mouse lemur*
	Cheirogaleus	*major* greater dwarf lemur *medius* fat-tailed dwarf lemur* *trichotis* hairy-eared dwarf lemur*
Subfamily Phanerinae	*Phaner*	*furcifer* forked-marked dwarf lemur*
Family **Lemuridae**	*Lemur*	*catta* ringed-tailed lemur *macaco* black lemur* *fulvus* brown lemur *mongoz* mongoose lemur* *coronatus* crowned lemur *rubriventer* red-bellied lemur
	Varecia	*variegatus* ruffed lemur
	Hapalemur	*griseus* grey gentle lemur* *simus* broad-nosed gentle lemur*
Family **Lepilemuridae**	*Lepilemur*	*dorsalis** *ruficaudatus* red-tailed sportive lemur* *edwardsi* Edward's sportive lemur *leucopus** *mustelinus* *microdon* *septentrionalis*
Family **Indriidae**	*Avahi*	*laniger* avahi or woolly lemur*
	Propithecus	*verreauxi* Verreaux's sifaka* *diadema* diademed sifaka*
	Indri	*indi* indris*
Superfamily **DAUBENTONIOIDEA**		
Family **Daubentoniidae**	*Daubentonia*	*madagascariensis* aye-aye*

SUBORDER: STREPSIRHINI

	GENERA	SPECIES
Superfamily **LORISOIDEA**		
Family **Lorisidae**		
Subfamily Lorisinae	*Loris*	*tardigradus* slender loris
	Nycticebus	*coucang* slow loris
	Arctocebus	*calabarensis* golden potto or angwantibo
	Perodicticus	*potto* potto
Subfamily Galaginae	*Galago*	*alleni* Allen's bushbaby *crassicaudatus* thick-tailed bushbaby *senegalensis* lesser bushbaby *inustus* *demidovii* Demidoff's dwarf galago *elegantulus* needle-nailed bushbaby

SUBORDER
HAPLORHINI

	GENERA	SPECIES	
Superfamily **TARSIOIDEA**			
Family **Tarsiidae**	*Tarsius*	*syrichta* Philippine tarsier* *bancanus* Horsfield's tarsier* *spectrum* spectral tarsier*	
Superfamily **CEBOIDEA**			
Family **Callitrichidae**	*Cebuella*	*pygmaea* pygmy marmoset	
	Callithrix	*argentata* silvery marmoset* *humeralifer* Santarem marmoset*	
		jacchus common marmoset *aurita* buffy-tufted-ear marmoset* *flaviceps* buffy-headed marmoset* *geoffroyi* white-fronted marmoset *pencillata* black-pencilled marmoset	*jacchus* group

SUBORDER: HAPLORHINI

	GENERA	SPECIES	
Subfamily Colobinae *continued*	*Simias*	*concolor* simakobu*	
	Rhinopithecus	*roxellanae* golden snub-nosed monkey*	
		avunculus Tonkin snub-nosed monkey	
	Pygathrix	*nemaeus* douc*	
Subfamily Cercopithecinae	*Cercopithecus*		
	Subgenus: *Cercopithecus*	*mitis* blue monkey	*mitis* group
		nictitans white-nosed guenon	*mitis* group
		albogularis Sykes' monkey	*mitis* group
		ascanius redtail	
		petaurista lesser white-nosed guenon	
		erythrogaster red-bellied guenon	
		cephus moustached monkey	*cephus* group
		erythrotis russet-eared guenon	*cephus* group
		mona mona monkey	*mona* group
		campbelli Campbell's guenon	*mona* group
		pogonias crowned mona	*mona* group
		wolfi Wolf's mona	*mona* group
		diana diana monkey	
		lhoesti L'Hoest's guenon	
		neglectus de Brazza's monkey	
		dryas	
		hamlyni owl-faced guenon	
		aethiops green or savannah monkey	
	Subgenus: *Allenopithecus*	*nigroviridis* Allen's swamp monkey	
	Miopithecus	*talapoin* talapoin	
	Erythrocebus	*patas* patas or hussar monkey	
	Cercocebus	*torquatus* smoky mangabey	*torquatus* group
		galeritus crested or agile mangabey*	*torquatus* group
		albigena grey-cheeked mangabey	*albigena* group
		aterrimus black mangabey	*albigena* group
	Papio		
	Subgenus: *Papio*	*hamadryas* sacred or hamadryas baboon	
		anubis olive baboon	savannah baboons
		cynocephalus yellow baboon	savannah baboons
		papio Guinea baboon	savannah baboons
		ursinus chacma baboon	savannah baboons
	Subgenus: *Mandrillus*	*sphinx* mandrill	forest baboons
		leucophaeus drill*	forest baboons
	Theropithecus	*gelada* gelada	

	GENERA	SPECIES	
Subfamily Cercopithecinae *continued*	*Macaca*	*silenus* lion-tailed macaque or wanderoo*	*silenus* group
		nemestrina pig-tailed macaque	*silenus* group
		pagensis Mentawai macaque*	*silenus* group
		tonkeana Tonkean macaque	*silenus* group
		maura Moor macaque	*silenus* group
		ochreata booted macaque	*silenus* group
		nigra Celebes black or crested macaque	*silenus* group
		sylvana Barbary macaque*	
		sinica toque macaque	*sinica* group
		radiata bonnet macaque	*sinica* group
		assamensis Assamese macaque	*sinica* group
		thibetana Père David's macaque	*sinica* group
		fascicularis long-tailed macaque or kera	*fascicularis* group
		cyclopis Taiwanese macaque	*fascicularis* group
		mulatta rhesus macaque	*fascicularis* group
		fuscata Japanese macaque	*fascicularis* group
		arctoides stump-tailed macaque	
Superfamily **HOMINOIDEA**			
Family **Hylobatidae**	*Hylobates*		
	Subgenus: *Hylobates*	*lar* lar or white-handed gibbon	
		pileatus capped gibbon*	
		moloch silvery gibbon*	
		muelleri Bornean or grey gibbon	
		agilis agile or dark-handed gibbon	
		hoolock hoolock gibbon	
		klossii Mentawai gibbon*	
	Subgenus: *Nomascus*	*concolor* concolor or crested gibbon*	
	Subgenus: *Symphalangus*	*syndactylus* siamang	
Family **Pongidae**	*Pongo*	*pygmaeus* orang-utan*	
	Gorilla	*gorilla* gorilla*	
	Pan	*troglodytes* chimpanzee*	
		paniscus pygmy chimpanzee or bonobo*	
Family **Hominidae**	*Homo*	*sapiens* human	

Glossary

Alpha male: The highest ranking male within a social group. He is dominant to all other animals in the group, and can displace them from valuable resources such as good feeding sites.

Beta male: A male who is subordinate to an alpha male.

Brachiation: A form of locomotion in which the animal travels by swinging underneath branches using its arms.

Caecotrophy: The re-ingestion of faeces. By this process, food is passed through the gut a second time so that more of it can be digested. This allows an animal to obtain nourishment from low quality foods.

Competitive exclusion: An animal, group of animals or species being denied access to a mutual resource by another animal, group of animals or species.

Diurnal primate: A primate which is active during the day.

Dominance hierarchy: The domination of some members of a group by others in a relatively orderly and stable manner. Except for the highest- and lowest-ranking animals, any given individual dominates one or more of the others and is dominated by one or more of them. The hierarchy is initiated and maintained by aggressive behaviour, although this may sometimes be subtle and indirect.

Ecology: The interactions of organisms with their environment and the scientific study of those interactions. The environment includes both the physical environment and also all plants and animals living in it.

Epiphytic plant: A plant growing on another plant but not parasitic upon it.

Evolution of the species: A gradual change in the genetic composition of a species population. This is brought about by differential mortality of and reproduction by individuals of different genetic compositions *i.e.* by the process of *natural selection*. Thus, the characteristics of the whole population gradually shift.

Exudate tapping: Obtaining gums and saps by biting, gnawing or scratching through the bark of a tree.

Folivorous primate: A primate whose diet mostly comprises leaves.

Forest canopy: The dense, continuous layer of vegetation in a forest composed of branches of trees and the leaves, flowers and fruits which they bear.

Fovea: The thinnest part of the retina of the eye. It is a region of very acute vision and colour perception.

Frugivorous primate: A primate whose diet mostly comprises fruit.

Genus (plural: Genera): A group of closely-related species. In the Linnaean system of nomenclature, the first name is that of the genus, the second that of the species. A single genus may contain one or many species.

Grooming: The cleaning of the body surface by licking, nibbling, scratching or picking with the fingers. In primates, this has both hygienic and social functions.

Gut microflora: The microbes found in the guts of animals. In some such as colobine monkeys and cattle, the microbes break down large molecules in food which the host animal cannot digest e.g. cellulose. This allows the animal to obtain energy from those foods which it would not otherwise be able to do.

Harem group: Social unit comprising a single adult male, several adult females and their mutual offspring. Also known as uni-male or one-male group.

Hibernation: Almost or completely continuous sleep during winter. The animal becomes torpid, its metabolism slows down, and its body temperature drops to approach that of its surroundings.

Hybridization: The production of offspring by the mating of parents from different species or subspecies.

Infanticide: The killing of infants. By killing an infant which is not his own, a male can then mate with and produce his own offspring with that infant's mother more quickly than if the original infant had survived. This may increase the male's reproductive success if competition for females is intense.

Invertebrates: Animals which do not have a backbone. This includes all insects, spiders and crabs.

Ischial callosities: Specialised pads on the bottom to facilitate long periods of sitting.

Linnaean classification: The system of classifying organisms initiated by Carl Linnaeus, an eighteenth-century Swedish naturalist. He established the binomial (two-named) system of classification, the first name being that of the genus, the second that of the species.

Montane forest: Forest growing at relatively high altitudes. It is characterised by smaller, more closely-packed trees than forest in the lowlands. Due to the high humidity at such altitudes, many montane forest trees are covered with mosses and lichens.

Multi-male group: Social unit comprising more than one fully adult male, several adult females and their offspring.

Natural selection: The differential survival and reproduction by individuals within a population. This results in the most well-adapted organisms contributing more offspring to the next generation. This is generally regarded as being the main driving force of evolution.

Nocturnal primate: A primate which is active at night.

Oestrus: Recurrent, restricted period of sexual 'heat' in females marked by a strong sexual urge around the time of ovulation.

Photosynthesis: The process by which plants synthesise carbohydrates (sugars and starch) from water and carbon dioxide, using energy from sunlight with the aid of the green pigment chlorophyll.

Plant defence chemicals: Compounds produced by plants to protect them from being eaten by animals. There are two main types of plant defence chemicals: (1) toxins such as strychnine or cyanide, and (2) compounds such as tannin and fibre which act as digestion inhibitors.

Pollination: The transfer of pollen from the anther (male part of a flower) to the stigma (female part of a flower) leading to the production of fertile seeds.

Race: See *subspecies*.

Rain forest: Evergreen forest comprising extremely tall trees of many different species and large numbers of climbing plants and epiphytes. Grows exclusively in warm areas with heavy rainfall throughout the year.

Riverine forest: Forest growing adjacent to or near a river. In the tropics, it is characterised by being extremely dense and productive and having large numbers of climbing plants.

Savannah: Tropical grassland with intermittent trees or patches of trees. This habitat is widespread throughout Africa.

Sexual dichromatism: A consistent difference in body or fur colour between the males and females of a species.

Sexual dimorphism: A consistent difference in body form between the males and females of a species apart from the basic functional portions of the sex organs.

Species: The basic lower unit of classification of organisms (see *Genus*). A species consists of a population or series of populations of closely-related and similar organisms. It is sometimes defined as consisting of individuals which are capable of breeding freely with one another under natural conditions.

Subspecies: A sub-division of a species. It is usually used to describe a geographical race: a population or series of populations occupying a particular geographical area and differing genetically from other races of the same species.

Tapetum lucidum: A reflective layer in the retina of the eye. It is present in certain nocturnal animals such as lorises, and increases the efficiency with which they can see at night.

Taxonomy: Study of the classification of living organisms according to their resemblences, differences and evolutionary history.

Territoriality: The exclusive use of an area by an animal or group of animals. This is achieved by repulsion in the form of defence or advertisement.

Ungulate: A member of the large and widely-distributed order of mammals characterised by the possession of hooves.

Uni-male group: See *harem group*.

Bibliography

About evolution:

Dawkins, Richard *The Selfish Gene*. Oxford and New York 1976.
A really modern guide to the way in which evolution works.

Leakey, Richard, & Roger Lewin *Origins*. London and New York 1977.
A well-illustrated guide to human evolution.

Wilson, E. O. *On Human Nature*. Cambridge, Mass., and London 1978.
A biologist's viewpoint on the evolution of human behaviour.

About primates:

Dixson, Alan *The Natural History of the Gorilla*. London and New York 1981.
A detailed but very readable description of what is known about gorillas.

Jolly, Alison *The Evolution of Primate Behavior*. New York and London 1972.
The best general textbook on primate behaviour (a revised and updated edition is in preparation).
A World Like Our Own. New Haven, Conn., and London 1980.
A superb account of the natural history of Madagascar.

MacKinnon, John *In Search of the Red Ape*. London and New York 1974.
One man's story of an adventurous study of the orang-utan.
The Ape Within Us. London and New York 1978.
Very readable descriptions of the nature of different ape species and what they tell us about our own nature.

Merfield, Fred *Gorillas Were My Neighbours*. London 1957.
An old-fashioned, but fascinating, account of gorillas in the Cameroons between the two world wars.

Morris, Desmond *The Naked Ape*. London and New York 1967.
A controversial look at man as a primate.

Morris, Ramona & Desmond *Men and Apes*. London and New York 1966.
A history of man's relationships with the other primates.

Napier, J. R. & P. H. *A Handbook of Living Primates*. London and New York 1967.
The best available source for really detailed information, although it is now rather out of date as far as ecology and behaviour are concerned.

Schaller, George *The Year of the Gorilla*. Chicago 1964; London 1965.
Pioneering work that showed that gorillas could safely be studied at close quarters in Central Africa.

van Lawick-Goodall, Jane *In the Shadow of Man*. London and Boston, Mass., 1971.
The famous chimpanzees of Tanzania's Gombe National Park, described by the woman who pioneered the research on them.

Whitten, Tony *The Gibbons of Siberut*. London 1982 (also available in the USA).
Life on one of Indonesia's remotest islands during a study of its lesser apes.

Illustration Sources

Anthro-Photo File, Cambridge, Mass.: p. 111 (below) (D. J. Chivers); p. 130 (above) (Hrdy); p. 158 (above) (J. Popp); p. 159 (below) (De Vore); p. 171 (D. J. Chivers); p. 187 (David Agee)

Ardea London: p. 36 (Liz & Tony Bomford); p. 40 (right) (Kenneth W. Fink); p. 41 (I. R. Beames); p. 48 (right) (Liz & Tony Bomford); p. 54 (Liz & Tony Bomford); p. 55 (P. Morris); p. 58 (Jean Paul Ferrero); p. 62 (I. R. Beames); p. 77 (right) (P. Morris); p. 83 (P. Morris); p. 87 (François Gohier); p. 102 (right) (François Gohier); p. 139 (Kenneth W. Fink); p. 150 (I. & L. Beames), p. 158 (below) (Clem Haagner); p. 159 (above) (Clem Haagner); p. 163 (Kenneth W. Fink); p. 194 (I. R. Beames)

Phil Agland: p. 155

Dr S. K. Bearder: p. 27, p. 63, pp. 24–25, p. 162

J. O. Caldecott: p. 130 (below), p. 174 (left)

A. G. Davies: p. 65

Dr G.W. H. Davison: p. 138

Robin Dunbar: p. 166, p. 167

Dr M. Kavanagh: p. 95 (left), p. 115, p. 127, p. 147 (above), p. 170 (left), p. 170 (right), p. 175, p. 182, p. 198 (top), p. 198 (centre), p. 198 (bottom), p. 202

Lysa Leyland: p. 147 (below), p. 154

Dr John MacKinnon: p. 59, p. 183, p. 186, p. 190

Foto Luiz Claudio Marigo, Rio de Janeiro, Brazil: p. 73, p. 76, p. 77 (left), p. 80 (above right), p. 82, p. 94, p. 99 (above), p. 99 (below), p. 102 (left), p. 106 (above), p. 106 (below), p. 110, p. 111 (above)

Dr Giuseppe Mazza, Monte Carlo: p. 80 (above left), p. 95 (right), p. 107, p. 131 (left), p. 174 (right), p. 179

Russell A. Mittermeier: p. 80 (below), p. 91, p. 114, p. 135 (left)

Dr J. J. Petter: p. 29 (WWF), p. 32, p. 33, p. 44 (above left), p. 44 (above right), p. 44 (below), p. 50

E. D. Starin: p. 119, p. 123, p. 142, p. 151

David Stevens: p. 131 (right)

Survival Anglia Ltd: frontispiece (Alan Root), p. 37 (Maurice Tibbles); p. 40 (left) (Liz & Tony Bomford); p. 48 (left) (Liz & Tony Bomford); p. 51 (Liz & Tony Bomford); p. 86 (Alan Root); p. 122 (Alan Root)

Dr R. L. Tilson: p. 11, p. 135 (right)

WWF Photo Library: p. 191 (Ian Redmond); p. 195 (Ian Redmond)

Andy Young: pp. 68–69

Index